Introduction to Innovation and Technology Transfer

The Artech House
Technology Management and
Professional Development Library

Bruce Elbert, *Series Editor*

Applying Total Quality Management to Systems Engineering, Joe Kasser

Engineer's and Manager's Guide to Winning Proposals, Donald V. Helgeson

Global High-Tech Marketing: An Introduction for Technical Managers and Engineers, Jules E. Kadish

Introduction to Innovation and Technology Transfer, Ian Cooke, Paul Mayes

Preparing and Delivering Effective Technical Presentations, David L. Adamy

Successful Marketing Startegy for High-Tech Firms, Eric Viardot

Survival in the Software Jungle, Mark Norris

For further information on these and other Artech House titles, contact:

Artech House
685 Canton Street
Norwood, MA 02062
617-769-9750
Fax: 617-769-6334
Telex: 951-659
email: artech@world.std.com

Artech House
Portland House, Stag Place
London SW1E 5XA England
+44 (0) 171-973-8077
Fax: +44 (0) 171-630-0166
Telex: 951-659
bookco@artech.demon.co.uk

Introduction to Innovation and Technology Transfer

Ian Cooke
Paul Mayes

Artech House, Inc.
Boston • London

Library of Congress Cataloging-in-Publication Data

Cooke, Ian (Ian Ernest), 1948-
   Introduction to innovation and technology transfer/Ian Cooke and Paul Mayes.
     p.   cm.
Includes bibliographical references and index.
ISBN 0-89006-832-1 (alk. paper)
1. Technology transfer. 2. Technological innovations. I. Mayes, Paul. II. Title.
T174.3.C65 1996
658.5′7—dc20   95-48910
                                                CIP

A catalogue record for this book is available from the British Library

© 1996 ARTECH HOUSE, INC.
685 Canton Street
Norwood, MA 02062

International Standard Book Number: 0-89006-832-1
Library of Congress Catalog Card Number: 95-48910

10 9 8 7 6 5 4 3 2 1

Contents

Preface

Introduction to Technology Transfer provides an overview and introduction to the topic of innovation and technology transfer. It introduces the subject and its many ramifications by the method of simplification. The topic is first broken down into its component parts and then reassembled, providing both a detailed and a holistic understanding. Throughout the book there are numerous examples drawn from a wide variety of industry and commerce, from many different countries. The book also addresses why innovation and technology transfer are necessary and what makes this process happen.

The content of the book will provide you with a good knowledge of the marketplace for innovation as it affects both the suppliers and the users of technology. It covers ways to recognize needs and opportunities for companies, the vast array of places to go for advice and help, and the evaluation and audit of technology. The book provides a basis for the understanding of the business and management issues affecting decision making, with regard to technology transfer opportunities including financial appraisal, market opportunities, and the establishment of market potential. We also give details about the legalities of intellectual property and securing the technology through patents. We examine the technology process as it would affect a company implementing a new product or service. These issues include manufacturing feasibility via product development, the establishment of good management systems, training, the launch of the product, and the monitoring and control of the new operation.

Innovation and technology transfer are a judicious blend of management and science, creativity with technology. All members of a production team, from fitter or cashier to marketing person, come within its ambit. The key person, though, is the one who makes the vital connection between the production or service process (portfolio) and the available technology. There are no absolute rules that, if followed, will guarantee successful innovation or technology transfer, but there are ways of optimizing the chances of making and implementing that crucial connection—these ways are the subject of this book.

We have one major specific objective—to develop *a better understanding of the total process of innovation and technology.*

The content is in nontechnical form so that it can be easily understood by anyone interested in the topic, irrespective of background. We hope it will prove particularly interesting for the following groups: owners, managers, research and development staff, trainers, and technical staff in manufacturing companies and research organizations; university students; industrial liaison staff in universities and colleges; staff in development and training organizations; fieldworkers engaged in company visit programs and market surveys; and people wishing to start their own business.

Chapter 1 and Chapter 2 provide an introduction and overview of innovation and technology transfer. There are numerous examples drawn from the past century of technological progress. Many of the issues relevant today were just as relevant 50 or 100 years ago. Some of the topics and examples examined in these chapters will be referred to in later chapters to be dealt with in more detail or from a different perspective. Here we also address the questions of why innovation and technology transfer are necessary and what makes it happen.

Chapter 3 to Chapter 6 will give you a good knowledge of the marketplace for innovation as it affects both the suppliers and the users of technology. They cover recognizing needs and opportunities for companies, as well as the vast array of places to go for advice and help in searching for, finding, evaluating, and auditing new technology.

Chapter 7 and Chapter 8 provide the basis for understanding the business and management issues affecting decision making with regard to technology-transfer opportunities. We discuss two essential requirements in the feasibility phase, the business plan and financial appraisal.

Chapter 9 deals with the very important and complex world of intellectual property. Included in the chapter are details on securing the technology through patents, licensing, and franchising.

Chapter 10 looks at the feasibility of the new product or service via prototyping and product development.

Chapter 11 to Chapter 15 examine the innovation and technology process as it will affect any company in implementing a new product, process, or service. These chapters look at practical examples of companies implementing a new product, and they review the issues raised by implementation. These issues include the establishment of good management systems, project management and the launch of the product, human resources and training, and the monitoring and control of the new operation.

If you are particularly interested in technology transfer and essentially topics that deal with the technical issues of innovation, these are covered in Chapters 1 through 6, Chapters 9 through 11, and Chapter 15 of the book. The aspects that are more inclined to business and management of innovation are covered in the other chapters.

Acknowledgments

We acknowledge that a course funded by the EU COMETT program and commissioned by the Regional Technology Centre North, Sunderland, England, forms the basis of the work in *Introduction to Innovation and Technology Transfer*. We are greatly indebted to Gordon Ollivere of the Regional Technology Centre North, who introduced us to the field of technology transfer some time ago and has kindly let us use his experience ever since.

We would also like to thank Terry Whatson of the Open University, John McManus of BioResearch Ireland, and Jenny Illingsworth for their help with Chapter 1, Chapter 9, and Chapters 10 and 11, respectively. Our special thanks are directed to Jeanette Crosby and Sarah White for their expert guidance on word processing.

We gratefully acknowledge the significant support given to us by our institutions in the preparation and writing of this book.

Finally, we are grateful to Julie Lancashire, Kate Hawes, and Laura Esterman at Artech House. Without their encouragement, optimism, and efforts, this book would never have appeared.

Writing a book consumes time that would have otherwise been devoted to family matters. For their understanding and provision of that necessary time, we wish to thank all the members of our families.

Our thanks go to all those, named and unnamed, who have contributed to this project.

Chapter 1

Introduction

1.1 INTRODUCTION

Many countries place a strong emphasis on the manufacture and sale of goods. The economic basis is that this is a means of creating wealth and thus a good standard of living for the inhabitants. There is in this a sort of circle of life. The industry provides work for people and this provides wages for the purchase of food and goods. If an industry can sell the goods abroad then this can create an influx of money to a country, because, basically, another country is paying the wages for the employees of that industry. This effectively means that there can be extra money for investment in the country producing the goods. The country is then creating wealth for its citizens. This is one of the aims of most governments. They desire to create or expand the commerce and industrial sector of their country to be greater than those of other countries and thus provide revenue. A country that produces most or all of its own goods and has many natural resources, such as the United States, is self sufficient. Even these nations wish to expand and grow economically, and to do that they look to sell their goods to other parts of the world. The use of this money enables the governments of these countries to put forth a policy of producing a high standard of living for the population. Hence there is the push to form new industries and to more effectively use the results of research and development to produce new products and services. There is, therefore, the impetus for industry and commerce to be continuously changing and to look for new ideas. This new push for change can be termed innovation and technology transfer. Yet what really are innovation and technology transfer? How can we use technology to our advantage? These are questions that this book aims to answer.

Manufacturing industry has always been in the business of using technology to produce merchandise. Whatever the period, at any given moment, the technology used in the manufacturing and in the merchandise would in most cases be regarded as "state of the art." Thus, the blacksmith's anvil, the radio valve, the steam

1

traction engine and even stone was state-of-the-art technology once. Yet state-of-the-art technology does not remain state of the art for long. The blacksmith's anvil was replaced by the hydraulic press, the electronic radio valve by the transistor, the steam traction engine by the diesel tractor, and stone by bronze. The relentless march of technology constantly threatens to leave a product or production process behind while it embraces new products or production processes. In essence, technology transfer is the process by which manufacturing industry keeps up with the march of technology. It is the process that ensures that production, once started, is maintained or—ideally—expanded. This way of looking at technology transfer appears to restrict it to the actual production process. Clearly, though, production cannot be started without appropriate facilities or finance and cannot be maintained or increased without a suitable market. This full process is the innovation cycle.

This book will therefore look at all of the aspects that affect an organization in its desire to progress and be innovative. We will consider in the various chapters the full innovation process from an original idea through to the launch of the new product or service. We should stress that the service industries must also be innovative to progress—it is not just the manufacturing sector that uses the innovation process.

The aim of this introduction chapter is to explain the foundations of innovation and technology transfer. A theme of the book is that innovation and technology transfer is an active process. It is not just a good idea that will come about in its own good time. It is essential, and it has to be worked at diligently and continuously if it is to be successful. The reasons why it is essential will become clear as you work through the book.

1.2 ROOTS OF TECHNOLOGY TRANSFER

As a way of introducing innovation and technology transfer, it is interesting to contrast the fortunes of two companies in the United Kingdom glass industry.

The method of glass production was brought to Britain by the Romans and Gauls. As far as one can ascertain from records, the first glass in England was made in the northeast region when, according to the Venerable Bede's *History of England* [1], Gauls, who hailed from what is now France, were brought in to glaze a monastery at Monkwearmouth, Sunderland, in 674 A.D. This can be recorded as a very early incidence of technology transfer with the Gauls bringing their skills and knowledge to another part of the world. Hence, a glass industry base was formed in the northeast of England. From that period on little happened to change the production process, which was essentially to manufacture glass in blocks. The next revolution in glass production occurred in the mid-19th Century, a revolution that made glass a commodity available to ordinary households. The firm of James Hartley, of Sunderland, discovered a new process of rolling glass that enabled the manufacture of large window sheets on the casting table. Their patent of 1838 was a fine example of technology exploitation. They tendered for the pavilion roof at the Great Exhibition of London and the firm grew to employ over 700 people. This was at a time when in many parts of the country, particularly the north of England, there was high unemployment. However, by the 1890s there had been further developments in glass technology, in particular in the furnaces—developments which Hartleys failed to recognize. It seemed as though they were relying on the patent and being somewhat complacent not thinking of progressing (i.e., being innovative). This was a costly failure for it ultimately led to the firm's collapse.

Hartley's was a world beater for some 50 years and its fortunes may have progressed for another 50 years had it continued to pursue new technology. That it did not pursue those technologies, that it rested on its patent and ultimately failed, attests to the inexorable march of technology and the importance for companies of being part of that march. The company joined with Alfred Wood, which produced colored glass. Now Hartley Wood specialize in the production of glass for the stained glass industry. It would seem that they have gone the full circle to return to the type of glass required in the region in 674 A.D.

In contrast, we can look at the fortunes of the Pilkington Glass Company. This company produced glass based on the principle invented by James Hartley. In the 1950s the company known as Pilkington Brothers, based in St. Helens in the northwest of England, was looking to improve the quality of the manufactured glass. This was to meet the market needs for shop windows, cars, and mirrors, where distortion-free glass was necessary. In the production process there was glass-to-roller contact; thus, the glass surfaces became marked. The glass had then to be ground and polished to produce the optical perfection required. An alternative to rolled-glass production was that for producing sheet glass. Sheet glass was made by drawing it vertically in a ribbon from a furnace. Sheet glass was cheaper to produce than polished glass because there was no need to grind and polish, but it tended to have more distortion. So the industry was looking to find a way to produce glass that had the best properties of the two methods. The process of float glass proved to be the required solution. The idea was that of Alastair Pilkington (no relation to the family owning the company), who apparently thought of the invention while washing dishes. The sight of a plate floating on water made him think that the same principle could apply to the manufacturing of glass. In the floating-glass process a continuous ribbon of glass moves out of the furnace and floats along the surface of a bath of molten tin. The ribbon is held at high enough temperatures and for a long enough time for the irregularities to melt. The surfaces become flat and parallel—because the surface of the molten tin is flat, the glass also becomes flat. The ribbon is then cooled down, while still on the molten tin, until the surfaces are hard enough for it to be taken out of the bath without rollers. Thus, there is no marking of the surface of the glass.

Very few inventions have an easy development to production. The Pilkington float-glass process took seven years to develop, 110,200 tons of scrap glass, and millions of pounds in investment to reach full production. It was a further 12 years before Pilkington broke even, but the subsequent payoff has been enormous in both business growth and royalty payments. The process is on license to organizations around the world, and licensing fees bring millions of pounds of revenue to the company each year.

It is noteworthy that the Hartley Wood glass company had a further crisis period in the 1980s. It is somewhat ironic that the company survived a takeover bid by the Pilkington Glass Company.

This story shows how companies cannot necessarily afford to depend on previous successes. They must continuously consider further development and innovation. It also shows how from a small but revolutionary idea—such as that of Alastair Pilkington—a company can innovate to produce huge amounts of profits for itself.

It is axiomatic that technology transfer begins with invention. After all, if something has not been invented, it cannot be "transferred" to the marketplace; it cannot be used in any products or processes. However, although invention may be the absolute beginning of technology transfer, the route by which that invention becomes a product (i.e., the relationship between an invention and its eventual application) is complex and incompletely understood. The complexity of this relationship and some of its contributing factors are illustrated in the case studies in Section 1.7 and in Chapter 2. First, we must first define what we mean by technology transfer.

1.3 TECHNOLOGY TRANSFER—A DEFINITION

A useful definition of technology transfer is: the technology-transfer process helps a manufacturing company more effectively use its human, physical, and capital resources by providing information or assistance,[1] which leads to improvements in its facility, equipment, manufacturing methods, management methods, or marketing methods.

This definition is all embracing, but a bit cumbersome; it may be accurate, but it provides you with no image of what technology transfer is. Indeed, it would be pointless at this stage to labor away at producing a snappy definition of technology transfer when there is the need for a clear understanding. What is innovation and what is technology transfer? First we will consider the component processes of technology transfer in turn. There are two reasons for doing this. The first is that there are a number of terms associated with technology transfer—such as invention, innovation, design, and marketing—that need differentiation not only from it, but also from each other. The second reason is that by knowing the components

---

1. The information or assistance should move the company closer to or beyond the current state of technology employed in the industry.

and understanding what the components are, it is possible to step back and see how they interact. Then one can see a better picture of their relationship to the innovation and technology-transfer process. That relationship will emerge in a very full way in Chapter 2.

1.4 WHAT IS INVENTION AND INNOVATION?

A very simple way of seeing how innovation works is to look at the example of the wheel. It seems incredible that the people of millions of years ago had the mental power to invent something that would be so useful for so long. They obviously saw a need for such a device and that the use of it would make certain tasks easier. Thus, there was the stage of invention. Someone then needed to devise how to really utilize the wheel; this is innovation. These ideas had to be turned into reality and other ideas had to be brought together to use the wheel as a practical implement; this is a design stage. For further developments to take place, the wheel needed promotion and the knowledge of the device had to be dissipated; this is diffusion. Thus, a simple schema of the innovation process would contain the components:

- Invention;
- Innovation;
- Design;
- Diffusion.

We will look in a little more detail at each component and then discuss the relationship binding them together.

1.4.1 Invention

Aluminum was invented in 1854 by a French professor of chemistry, H. Sainte-Claire Deville. This new metal was light, malleable, ductile, and not easily corroded. Despite these useful attributes, there were no obvious immediate uses for this new metal. Nevertheless, the Academie des Sciences allocated 2,000 francs for the furtherance of Deville's work, as they were convinced of the metal's potential. However, the potential of aluminum was not to be realized for almost 100 years. This was not until after the Second World War, when it was used in the form of alloys for domestic utensils, masts for sailboats, window frames, aircraft fuselage, and wings and coachwork. It was then that aluminum found its niche in the marketplace.

A similar story can be told of the laser device. The laser, or at least the theory on which it is based, was demonstrated by Einstein in 1917. It took some 43 years before the first working prototype gas discharge laser was made. It, too, remained a scientific curiosity for several years before becoming an essential component of optical communication. Another feature of present day optical communication is the property of internal reflection. This property traps light within a tube, even if the tube is transparent. The more highly polished the internal surfaces of the tube are, the more they reflect the light and the better the transmission of light. William Wheeler was aware of this and saw the potential of shining light around corners. He patented his optical transmission system in 1881 (some 90 years before fiber optics took off), though it never reached the production phase.

There are of course inventions of immediate appeal. One such was x-rays. These were discovered by Wilhelm Röntgen in 1895 and by the turn of the 20th Century x-ray tubes had been manufactured to serve the medical profession. This example is an exception. More often there is a time delay measured in decades between invention and application.

The time delay between invention and application begs one of the fundamental questions about technology transfer. Why do inventions take so long to reach the marketplace? Put another way, the question becomes: what factors govern the delay between invention and application? This is very important in the present age because speed to market is one factor that for some companies is proving the difference between success and failure. The first one onto the market with a new product clearly has a competitive advantage over rival firms for a while. Why do some inventions survive for long periods while others fade away? We all use the QWERTY system for the computer keyboard that was invented in the late 1800s. How has it managed to remain the standard when logic dictates that there should be better layouts for typing? People have invented other keyboard systems, but none have been successful. The best known alternative, Dvorak simplified keyboard (DSK), invented by August Dvorak, a professor of education at the University of Washington, was introduced in 1932. In the 1940s the records for speed typing and quickness in training typing were held by those using DSK, not QWERTY, typewriters. Yet the DSK never became popular and August Dvorak died in 1975, no doubt a disappointed inventor.

There turn out to be a great number of other important factors that prove to be barriers to innovation or cause delay in the development of an idea. You will encounter these throughout the book.

Some inventions are forced. This was so when a major research effort, backed by government money and resources, was put into radar in the early 1940s. The idea and (unworkable) design had been patented as long ago as 1914, but it was not until radar was seen as necessary that sufficient effort was put into its development. In the case of radar, the need preceded the product.

In a similar vein, the need for the transistor existed long before its invention on December 23, 1943. The valve, which it was to replace, was bulky, fragile, consumed a lot of power, had a restricted life, and was thus inherently unreliable. Any solid-state amplifier that could overcome some, or all, of these problems was a certain winner. Bell Laboratories made a huge investment to provide the right environment for research into the physics of materials. The reward was the invention and patent of the transistor (in 1948), but it was not until 1951 that manufacture of the point-contact transistor started in earnest.

The invention of fiber optics reveals just how complex the path to invention can be. As already stated, the concept of transmitting light through tubes was first used in a patent in 1881. Charles Boys had made long uniform fibers of quartz as small as 0.25 micrometers in diameter, whose properties he discussed in 1887. John Logie Baird had bundled glass fibers together to make cables in 1927. Van Hell improved the optical properties of bare glass fiber in 1952 by coating it. Alex Reeves was trying to use light modulation in communication by guiding the light output of hollow cathode-discharge lamps along designated paths in the late 1950s. The laser was invented in 1960. Charles Kao and George Hockham identified impurities in glass as the main causes of signal attenuation and wrote in 1966 that losses of only 20 dB per km were achievable, a figure actually achieved by Robert Maurer in 1970. Communication by optic fiber emerged from these various, separate events.

These examples illustrate that the inventor is part of a complex web of events. There is the experience and knowledge of the inventor. There is the intercourse between the inventor and colleagues, friends, rivals, and others. There is the current state of the market, with which the inventor is familiar; there are millions of previous inventions. The inventor latches onto the particular problem that is of interest and synthesizes a solution from his or her particular position in the web.

It has just been stated that the inventor latches on to a particular problem. Inventors not only recognize that a problem exists (e.g., how to use an engine to move people), but determine to tackle it. This determination is the first of five identifiable stages in the innovation process. A period of preparation follows, in which the problem is considered in more detail and various solutions are attempted. Preparation is followed by the third stage, incubation, which—after an uncertain length of time—results in illumination, the creative insight. Finally, the idea is worked up into tangible solution, verification.

It is the fourth stage, illumination, that is usually regarded as the most important, but it will not happen without the previous three stages. It is at the fourth stage that "chance" observation often occurs, too. It is not chance observation at all, but rather the recognition of the significance of an event. There may be many years between the first and last of these stages.

One measure of the frequency of invention is the patents list. A company with an important invention will patent it to secure and exploit it (see Chapter 9). Only three European companies feature in a list [2] of the 25 companies with the most U.S. patents in 1991; Philips, Siemens, and Hoechst. The other companies are either from Japan or the U.S. The top company, Toshiba, filed 1,156 patents in 1991. Hence the present push by the European Union (EU) to develop research and produce more benefits from this research (see Chapter 4).

In all of the examples considered so far, a great deal of development was necessary after the initial invention to produce a viable product. In the case of the transistor, it was not until 1954, 11 years after the invention, that Texas Instruments produced the first silicon transistor of wide application. The period of development is characterized by innovation itself, the second parameter of the innovation process.

1.4.2 Innovation

Once a product has appeared on the market it is possible to develop that product so that it is easier to manufacture or is more efficient and reliable. Thus, we distinguish between invention—the original conception—and production and innovation—the development and change of an existing product.

The triode or R-valve was invented by De Forrest in 1906 and was the valve on which the radio industry was built. It remained little changed for 15 years despite being bulky, expensive ($3, a not inconsiderable sum in 1915), and unreliable.

Stanley Mullard recognized the inadequacies of the prevailing valves and made three changes. First, he improved its reliability. He made the electrode assembly more rigid by mounting it vertically instead of horizontally. Second, he reduced loses due to breakage in transit by making the glass envelope cylindrical instead of spherical. Third, he streamlined production. Using relatively unskilled labor, he achieved low rejection rates and a consistent product. His ORA valve performed no better than other R-valves but was half the price and more reliable. These three innovations were sufficient to increase the profit of the Mullard Radio Valves Company tenfold, from £3,000 in 1921 to £33,000 in 1922. They also established the company as the market leader.

Once the 15 shilling (approximately one dollar) Mullard ORA valve was on the market, it became the target to aim at for other valve manufacturers. They did this not only by reducing their prices, but also by developing a more energy-efficient valve. In the 1920s, domestic radio valves used direct current from an accumulator or battery. The R-valves consumed a lot of power, so the accumulator needed to be recharged fairly frequently, which was both annoying and expensive. (Home users had to take their bulky accumulators to special recharging points, often in local bicycle shops, as frequently as once a week.) Thus, when "dull" emitters became available, which consumed one quarter to one eighth of the power of "bright" emitters, they soon captured the R-valve market. This particular story of relentless innovation, by which market leaders are continually superseded by better or cheaper products, could be continued right up to the eventual obsolescence of the valve with the advent of the transistor and beyond. Indeed, product competition is familiar to all.

Neither is continuous product innovation such as this anything new. All artifacts from bows and arrows and cartwheels to transistors have been improved over the years. The difference between then and now is twofold. First, there is the timescale over which improvements happen. The cartwheel was improved to dish-shaped perfection over hundreds of years; it evolved through countless numerous minor alterations. The valve on the other hand went through all of its changes in forty years. In the consumer electronics industry, model life has fallen from 1.5 years in 1980 to under one year in 1989. The second difference between earlier times and the present day is in terms of conscious design. No one sat down at a drawing board and decided to design a cartwheel. How it performed dictated its evolution: good performers were used as the model for further production, poor performers were scrapped. However, the intention today is to bring out a product that is an excellent performer from the outset. To achieve this, conscious design is necessary and so design is the third parameter of the innovation process.

1.4.3 Design

Roberts Radio provides a good example of the three main design processes. The intention of the founders of Roberts Radio was simply to produce the best radio. Their first task was to ensure that the electrical components were of high specification and that the circuit design was the best possible. These are features of engineering design. The end user need have no knowledge of these features, except through their performance. However, the end user does interact with the various controls, tuning knobs, controlling volume, and so on. What controls there should be and how they are arranged are features of technical design. Finally, the product should be attractive and something with which the user wants to be associated. The color, texture, material, and shape of the overall product are features of aesthetic or fashion design. Roberts Radio produced three radios a week in 1935, increasing to 10 per week in 1936, with much of their initial output being sold at a premium to the Harrods exclusive department store.

The alternative to exclusivity is mass production, and the early radio world provides an example of the role of chance and "gut reaction" in commercial success. In the spring of 1930, Michael Lipman, a London representative of the German electrical group AEG, decided he would like a day at the seaside. He arranged to visit a company of which he had never heard—E. K. Cole. (Ekco Radio), which had inquired earlier about some special insulation. He finished his business with the buyer, who then asked if he had anything that might be of interest for the mass production of radio sets. As it happened, Lipman had a bakelite radio cabinet in the back of his car in which he had unsuccessfully tried to interest several British radio manufacturers. The manager of HMV, one of the largest radio manufacturers at the time, saw no prospect for a product so totally alien to the high-quality wooden cabinets that its dealers and customers expected. Other companies had been reluctant to agree to the 10,000 minimum order. The Ekco buyer liked the cabinet, agreed to terms, and stole the show at the radio exhibition in the autumn of the same year. Fashion design had been sufficient to capture a large part of the radio market for Ekco.

1.4.4 Diffusion

So far the terms *invention, innovation,* and *design* have been considered. One further term, *diffusion,* whereby the product gets into the market place, completes the list of factors that comprise the innovation process. Diffusion is everything that is involved in the promotion and sale of a product and is often simply referred to as marketing. However, marketing is that part of the diffusion process over which the company has control, such as advertising, promotion, and pricing. The other part of the diffusion process, over which the company has no control but which is still important to market penetration, is essentially promotion by use or publication. Examples would include a customer telling a friend about the benefits of a particular product or an inquiry about a product that has arisen because someone has seen a particular product in use, or the publication of the details of the product in a research paper. Diffusion is very important because it comprises of the knowledge of a product or service. Companies may want to search for a new product so they will have to look for sources where details of products are available. Knowledge of a product means it can be copied so companies will want to protect the products using patents. Companies are particularly concerned about the marketing part of diffusion, but many other important aspects of diffusion will be discussed in this book.

1.4.5 The Relationship Between Invention, Innovation, Design, and Diffusion

The four principle components or parameters of the innovation process are invention, innovation, design, and diffusion, but what is the relationship between them? One way of illustrating the relationship is as a linear sequence in which invention, because it occurs first, appears on the left and diffusion, because it occurs last, appears on the right:

Invention → Innovation → Design → Diffusion

The sequence is, of course, much more complex than this diagram would infer. First, there is feedback from each stage to earlier stages in the sequence. Second, the real-life sequence does not have an obvious beginning nor an obvious end. Invention does not occur at one moment in time but is continual. Hence innovation has become a term to describe the whole process, embracing all factors rather than being a just a component, and technology transfer is the means by which you achieve innovation. Figure 1.1 accommodates these criticisms and recognizes that the interaction between the components is a continuous process.

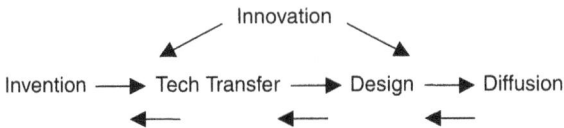

Figure 1.1 Interaction between the components of the innovation process.

In Chapter 2 we will consider again the innovation process. You will see that the process is far more complex than even this representation. The components in Figure 1.1 can be broken down to give a more explicit picture of how a company progresses from an idea to the actually marketing of a new product. We can now also give a simpler definition of technology transfer: The transaction between changing technology and invention (innovation), production, and diffusion.

There is also a simple definition of innovation: The exploitation of new ideas.

The preference for most organizations would be that this is the successful exploitation of new ideas.

1.5 INNOVATE OR LIQUIDATE

A worldwide study by Arthur D. Little [3] found that managing directors, or chief executive officers, believed that innovation was becoming more important: 97% of those surveyed in Japan, 87% in Europe, and 92% in the U.S. The reasons for this belief can be found in some of the previous sections of this chapter. The separate innovations of Mullard and Ekco gave them a considerable market edge and increased sales. Increased sales ensure company survival and profitability, which are obviously the key reasons behind innovation. Most companies believe that to survive or increase profits, they must have a competitive advantage over their rivals. To do this they feel they must be innovative. The U.S. company 3M has always had a policy of encouraging creativity. They have a renowned tolerance for trying new ideas, believing that unrestrained creative thinking will always pay off in the end. The result is that the company produces new products generally faster and better than most other organizations.

One of the most inventive companies in the world is Sony of Japan. They have a policy of being ahead of competitors, which in itself creates its own pressures.

As Akio Morita, the CEO of Sony, stated in his book [4]:

We were bringing out some products that had never been marketed before—never made before, such as transistorized radios and solid-state TVs—and were beginning to get a reputation as a pioneer...We would produce a new product; the giants of the industry would wait to see if it was successful; and then, if it was, they would rush a similar one onto the market to take advantage of our efforts. This is the way it has developed over the years; we have always had to be out in front.

So Sony has created many new products such as transistorized TVs, handheld flat TVs, video cassette recorders for home use, Trinitron color TV systems, 3.5-in floppy disks, handheld video cameras, filmless cameras, and compact discs. Yet even Sony almost didn't produce a product that has proved to be revolutionary and bought by millions of people around the world—the Sony Walkman cassette player.

The CEO of Sony had great difficulty pushing this idea through to development. The initial costs of production were higher than expected and the cost of each unit to the public was also initially high. They persisted, though, and brought down the price. The result was that very soon after entering the market, the company could hardly keep pace with demand and had to design new automated machinery to handle the flood of orders. The idea for the Walkman is in a way typical of how a conversation between two people could eventually produce such a successful profit-making product.

The idea took shape when one of the founders of Sony, Masaru Ibuka, went into the office of the CEO with a portable stereo tape recorder and a pair of standard-sized headphones. Ibuka was unhappy and complained about the weight of the system. The CEO asked him what was on his mind and then Ibuka explained [4], "I like to listen to music, but I don't want to disturb others. I can't sit there by my stereo all day. This is my solution—I take the music with me. But it's too heavy."

A point that Morita stresses in his book is that he does not believe that any amount of market research could have told them the extent that the Sony Walkman would be successful and widely imitated. Yet this small portable product has changed the way millions of people around the world of all ages listen to music. Finally, it is interesting to look at a philosophical statement from the original prospectus of Sony [4]:

If it was possible to establish conditions where persons could become united with a firm spirit of teamwork and exercise to their heart's desire their technological capacity, then such an organization could bring untold pleasure and untold benefits.

The essence of this story is that companies must innovate to climb ahead of their competitors, but there is no magic formula for devising the winning product.

A definitive strategy may work for one company, but luck and bright ideas (such as those of Alastair Pilkington mentioned in Section 1.2 and of Ibuka here) defy any laws one could devise for successful innovation.

However, there are several other reasons for innovation, all of which have to do with the inevitability of change.

1.5.1 Additional Reasons for Innovation

1.5.1.1 Competition

Consider a motorcycle company that has produced virtually the same large motorbike for 25 years. The company has a loyal, if long-suffering, clientele and is convinced that a motorbike that has withstood the test of time can continue to do so. The company knows that the engine is prone to oil leaks and regards the ticking and clicking noises as part of the bike's character and so does nothing about them. The company also believes that foreign producers of small motorbikes will not produce a large motorbike to compete. What would be the effect on the company of someone else importing a quiet, leak-proof, large motorbike?

The scenario just described was more or less what happened to the Triumph motorcycles in the U.K. in 1965. The Japanese company, Honda, which had been producing small motorcycles for a number of years brought out a large (750cc) model. Within a year Triumph had responded with the three cylinder Trident, but the engine had been a rushed design that was stuck into a frame designed for another engine. Neither the Trident nor Triumph lasted very long after that.

What is striking about the Triumph motorcycle story is that many good innovations had been devised, but not pursued. An engine, very similar to those later produced by Honda, was designed at Triumph in the 1960s. The Trident engine was first conceived in 1960, but not produced. The spur of competition forced the change—but too late for the company.

1.5.1.2 Science and Technology

The transistor was a considerable breakthrough and forced whole companies to redesign circuits to accommodate the new unit. While the impact was strongest on radio and television manufacturers, it wasn't long before most industries were affected. Texas Instruments, which had patented the silicon transistor in 1954, soon introduced another innovation that had just as dramatic an effect. The integrated circuit, whereby the functions of several discrete components are performed within a single chip of semiconductor material, was patented, again by Texas Instruments, in 1959. Suddenly the whole electronics industry had to change again.

The point of this potted history of the electronics industry is not to reveal who invented what, but rather to demonstrate that changes in available technology inevitably led to changes in the industries that used those technologies. Furthermore, the changes in the available electronics also led to changes in manufacturing processes. Were the changes inevitable? Was the invention and development of the integrated circuit the primary reason for the widespread adoption of the integrated circuit in electronics industries? These questions encapsulate the notion that technological change can be the result of science push and thus changes in the science and technology base lead to product innovation (i.e., technology push).

1.5.1.3 The Market

Customer feedback on products has always had an influence on innovation leading to product modification. Throughout the 1980s, customer awareness of environmental issues led to the development of environmentally friendly products and processes. Recycling paper into toilet paper rolls and the air canister, an aerosol that uses air as the propellant instead of volatile gasses such as chlorinated flurocarbons (CFCs) and propane, are direct results of changes in the market.

1.5.1.4 Legislation

Government legislation can force innovation in a variety of ways. Such legislation may insist on a particular safety standard being met for a seat belt or crash helmet. It may insist that a particular dye or flavoring be replaced. On a larger scale it may set rigorous environmental targets such as those on industrial discharge into rivers or the atmosphere. All of these can lead to innovations in manufacturing procedures and products. At the Third International Conference on the Protection of the North Sea in 1987, the U.K. government undertook to end sea dumping of industrial wastes by mid-1993. Stirling Organics was affected by this legislation because it dumped waste from the production of paracetemol into the North Sea. It developed the process of wet air oxidation to treat its waste as a direct result.

Changes in economic policy can spur the need to find cheaper or more efficient production processes. They may even trigger the search for alternative raw materials.

1.5.1.5 Human Nature

The proximate causes of innovation discussed above all appear to originate from outside the company. Within the company, at all levels, two fundamental characteristics of human nature can be harnessed. These are curiosity and laziness. Curiosity, the "what happens if I do this?" mentality, is difficult to satisfy outside a research laboratory or university, but the questions should be received and considered. Laziness, characterized by the "there must be an easier way to do this" approach could be more easily harnessed. Both require good communication within the company.

1.5.1.6 Innovation as a Policy

While innovation does not guarantee success, lack of it will ultimately lead to failure. Thus, innovation and invention are integral components of multinational companies such as Imperial Chemical Industries (ICI) and Philips. Indeed, one company, Nissan, has taken innovation a stage further to the realms of imagination. Each year the company holds a competition for all employees to come up with a novel form of transport. It isn't clear exactly what the rules are for winning the competition, but a prototype of the winning entry is made. Two such winners were the powered pogo stick and the "scrubbing brush" car. The scrubbing brush car was based on the principle that a brush placed on a vibrating surface (e.g., a washing machine on the spin cycle) will move in the opposite direction to the angle of the bristles. The design generated the vibration from within the vehicle. Speed was controlled by the speed of the vibration, and direction was controlled by altering the angle of the supporting struts (bristles). Clearly, such a vehicle has no practical application or market, so what is in it for the company?

Morale is one important consideration. A second consideration that is more relevant to the present course is that any novel design throws up numerous technical problems that have to be solved for the prototype. In solving those problems engineers may well see solutions to other technical problems. For instance, did solving the vibration problems of the scrubbing brush car lead to a better vibration control in other products?

The U.K.-based Cookson Group also has innovation as an integral component of its commercial strategy. In 1984 the company began to develop liquid-metal dispensing in its Fry's Metal subsidiary. It took the time-honored "lost wax" foundry process for precision casting and replaced wax with a low-melting-point tin alloy tailored to the demands of high-speed injection molding. It did this in close collaboration with its customers, who wanted high precision on the internal surfaces of plastic moldings. The upshot is not only a new material but a novel £300,000 production system known as Fry's multipurpose liquid metal dispenser.

Where innovation is policy, change is inevitable and in part is driven from within the company. The company is not simply waiting to respond to one or another of the external proximal causes, but is itself the vehicle for change.

1.5.2 Benefits of Innovation

If there is the need to innovate to survive (which, of course, is in itself a major benefit), what other gains exist to a company through innovation? The main benefits are:

- Competitive advantage;
- Increased market share;
- Higher growth rates.

These, then, lead to a significant indicator for a company—increased profit.

The former president of the Board of Trade in the U.K., Michael Heseltine, stated while in that post that "In itself R & D does not create a successful economy. It is successful exploitation that is the key to competitiveness and wealth creation" [5].

It is certainly the hope and desire of governments and groups such as the EU that research can and will play a vital role in promoting growth and improving the quality of life. This, they state, will be achieved by increasing knowledge and skills and by virtue of the economic spin-offs of research. The development of a new product that creates such wealth though can be analogized to be somewhat like a fairy tale in that the invention relates to "Once upon a time..." while the product relates to "...happily ever after."

If we consider product survival, then over the centuries the length of successful survival has shortened dramatically as Table 1.1 shows:

Table 1.1
Product Survival

Period	Length of Product Survival
C13–C19	Century
C19–1940	Decades
1940–Present	Years
Present–Future	Months

In their book on the history of the semiconductor industry, Braun and Mac-Donald write [6]:

As the first commercial representative of the new electronics, the germanium point contact transistor was not an impressive creation. Certainly, it marked an impressive breakthrough, but it was a manufacturer's nightmare. It was difficult to produce reliable transistors and even harder to produce them with identical characteristics; life expectancy was uncertain and the transistors seemed inclined to deteriorate rapidly under temperature and humidity conditions which were far from extreme. When these manufacturing difficulties were considered alongside the technical limitations..., its inability to handle high frequencies and large currents and a tendency to produce a great deal of noise, the transistor seemed a troublesome and unattractive product.

The first reliable silicon transistor was not announced until May 1954, some seven years after the transistor was invented. Thus, the transistor was not generally regarded as an electronic miracle in the early fifties.

The point of this example in this context is twofold. First, the existence of the transistor was not sufficient reason for manufacturers to embrace it; it had to prove it could achieve its suggested advantages. Second, some people (e.g., those at Bell, Texas Instruments, and Philco) must have continued to believe in the potential of the transistor, despite its early setbacks. These people, the ones with vision, tenacity, and belief are not necessarily those actually doing any research or invention, but they are the ones that make things happen. They glorify in the name of the product champion. This again shows the complexity of the innovation process, particularly if a company is using it to stay afloat. One can, though, ascertain from the many examples described that perseverance is an essential factor if you really do believe in the product and want to get it developed. The difficult art is getting the correct balance between the expected development costs—and the fears of such—and the expected income from sales. This is the fight between short-term thinking (the development costs), which often affect accountants and many managers, and the long-term profitability (the income from sales). This is where the product champion is required. The product champion must battle for the development and production of the new idea to make, hopefully, huge profits in the future.

Two other examples that were introduced earlier illustrate the limitation of the science-push model of innovation. Aluminum was invented a long time before it had any applications. If science push was important, aluminum products would have appeared much sooner after its invention. The second example is of the scrubbing-brush car mentioned in Section 1.5.1.6. Here was an idea of impressive originality, yet it led absolutely nowhere.

If we look back at the development from the vacuum tube to the semiconductor, it is interesting to see how the manufacturing companies of these products have progressed, moved to other products, diversified, or gone out of business. Table 1.2 lists the top ten manufacturing companies in this sector in each period. You might like to study the table and think about how these companies performed over the period and whether they are in existence today.

Two examples that were introduced earlier illustrate the limitation of the science-push model of innovation. Aluminum was invented a long time before it had any applications. If science push was important, aluminum products would have appeared much sooner after its invention. The second example is of the scrubbing-brush car mentioned in Section 1.5.1.6. Here was an idea of impressive originality, yet it led absolutely nowhere.

Table 1.2 shows how even when products were supposed to last for years, companies needed to innovate and be aware of change to survive. In the present period, when customers are constantly looking for better quality and something new, it is clearly essential that companies continually review their processes. There is the ever-present battle to stay ahead of the competitors and to do this they will in many cases need to be innovative. Thus, there is the reason to constantly emphasize the phrase or slogan for companies "Innovate or Liquidate!"

Table 1.2
Manufacturing Companies 1945–1982

Rank	1950 (Vacuum Tubes)	1955 (Transistor)	1965 (Semi-conductor)	1975 (ICs)	1980 (LSIs)	1982 (VSLIs)
1	RCA	Hughes	TI	TI	TI	Motorola
2	Sylvania	Transitron	Fairchild	Fairchild	Motorola	TI
3	GE	Philco	Motorola	National	National	NEC
4	Raytheon	Sylvania	GI	Intel	Intel	Hitachi
5	Westinghouse	TI	GE	Motorola	NEC	National
6	Amperex	GE	RCA	Rockwell	Fairchild	Toshiba
7	Nat Video	RCA	Sprague	GI	Hitachi	Intel
8	Rawland	Westinghouse	Philco/Ford	RCA	Signetics	Philips
9	Eimac	Motorola	Transitron	Philips	Mostek	Fujitsu
10	Lansdale	Clevite	Raytheon	Amer'Micro	Toshiba	Fairchild

1.6 INFORMATION

Inventors and innovators are able to do what they do primarily because of the information available to them. Without information, innovation is just not possible. Optimizing the flow of information is at the heart of innovation and technology transfer, and we will deal with it in more detail in Chapter 3.

How is it possible to influence the flow of information? This is a key question, which is easier to answer when information is separated into three distinct types:

- Knowledge-based (ideas, thoughts, plans, patents, and theories) information;
- Skills-based information;
- Equipment-based information.

1.6.1 Knowledge-Based Information

This type of information can be conveyed in print through technical and learned journals, scientific magazines and patents, and by word of mouth at conferences, at learned societies, and in discussion with colleagues and acquaintances. It may feature on the occasional television or radio program and, of course, it is possible to take courses to bring the knowledge base in a particular field up to date. In many ways this type of information is abstract or academic. An example is the idea that the transmission of light through a glass fiber is limited by the impurities in the glass, rather than by any intrinsic properties of the glass. This information, first published in a learned journal in 1966, led glass manufacturers to improve the quality of glass that paved the way for fiber optics.

1.6.2 Skills-Based Information

This type of information is acquired by doing something—it is hands-on experience. It can be conveyed by watching someone do something, either on video or live, but is best conveyed by participation. Such skills can be learned on a course or by sending people to other institutions where the appropriate skills are practiced. Alternatively, skilled personnel can be exchanged or bought in. An example of this type of knowledge would be Mullard's improvements to the R-valve, which were based on his knowledge and experience of the production process.

1.6.3 Equipment-Based Knowledge

This type of knowledge is conveyed via products. Knowing what machine tools are available and what they can do, what merchandise is available in the market place and what features it has. This type of knowledge is conveyed via trade magazines and conventions. It is also conveyed by sales representatives, direct mail, and advertising. Another information channel is through contacts in other companies or acquaintances who move in other spheres, both of which allow you to know what equipment other companies are using. An example of equipment-based knowledge is waterjet cutting. A very fine jet of water at high pressure is able to cut almost anything from titanium, copper, and stainless steel through to ceramics, glass, and concrete. There is no heat or dust or deformation and the technique can cut in any direction. Knowing and using such equipment allowed much more efficient use of the titanium used in the manufacture of hip joints.

Given that there are three types of information, it is possible to consider the flow for each of them. Fortunately, a single distinction—between formal and informal information sources—is appropriate for them all and this distinction enables a source of control to be seen. Formal information sources are those that are consciously controlled, for example, subscription to a trade magazine, membership of a learned society, affiliation to a regional technology center, and engagement of a consultant. The formal information sources in this list are all outside the company. Thus, they would be used to get information into the company. However, there is also information within the company, and it is important not only to recognize it, but also to use it. (How does the person on the shop floor, who has come up with a way of improving the use of a piece of equipment, get that information to the decision makers in the company?) Familiar formal mechanisms for conveying information within the company are meetings, newsletters, movement of personnel, and training (staff development).

The informal flows of information are, by definition, unplanned, random meetings between people and between people and other things (i.e., equipment or writings). Such flows could be over-the-fence conversations, forced (and initially unwanted) groupings at conferences or on trains, or chats at lunch time. (Mullard's attempt to produce a dull emitter was saved by an acquaintance at a rival factory, Osram. The acquaintance told Mullard that a little magnesium vaporized inside the valve, forming a deposit on the glass that absorbed residual gasses and prevented poisoning of the filament. Mullard's informal network was very close to industrial espionage!)

Formal information flows are much easier to control than informal flows. This is because formal information flows are consciously decided (e.g., going to a conference, subscribing to a magazine, belonging to a business association). Furthermore, many of these formal contacts would provide information on a regular basis. Informal information flows are no less important, but by their very nature are much harder to control.

However, and this is the salient point here, it is only individuals who use information; the company cannot use the information. A company cannot connect two units of information—two ideas—together, only an individual can do that. Even in research teams in large companies, where a group of individuals is working on the same problem, the solution will occur to an individual, not to "the team." In these circumstances, two individuals may come up with the same solution at the same time, but ultimately they arrived at their solutions as individuals.

The importance of individuals when considering the management of information flows cannot be overstated. Individuals must have access to information. If they do have access to information then they can make the crucial connection between a process and a new machine, an idea and an improved product or service. In short, invention and innovation are possible. Yet these are only a part of the innovation and technology-transfer process—a crucial and necessary part, definitely, but only a part. One of the next stages in the process would be to ensure that the connection is made known to the relevant decision makers within the company and to potential product champions. These people are most likely to be line managers; in this context though, they are information managers, and information managers must also be aware of market information so that they too can make necessary connections.

1.7   CASE STUDY—SEABAIT

Seabait of Northumberland, England, is a relatively new company that was incorporated in October 1985. The company produces bait for sea anglers. This might seem a strange product, but the formation and establishment of the company is a classic example of the innovation and technology-transfer process. The case study will show how complex and difficult it is to progress from an idea through to a product then to commercial sales. The case study will lead us into the formulation of a new set of tasks for the innovation and technology-transfer process in Chapter 2. The case study will also be used for illustration and consolidation of points throughout the book.

Peter Cowin embarked on a degree course at the University of Newcastle with the intention of using his newly gained knowledge in the fish-farming industry at some stage in the future. The expertise at the university was not, though, in fish but in marine worms. Peter decided, therefore, to use his course project work time to look at the viability of the commercial production of rag worms. Dr. Peter Olive, a senior lecturer at the university, had already carried out some studies in bait digging and was aware of the commercial possibilities for rag worms. Peter Olive was the obvious person to be a supervisor for Peter Cowin's projects. The undergraduate projects really could only be considered a beginning of the investigation. It was clear that there was enough work in this area for the research to be continued to a Doctor of Philosophy program.

It was also clear by the second year of the Ph.D. that the commercial production of rag worms was viable. In 1984, the university had set up the Newcastle University Technology Centre (NTC), which had as one of its objectives the transfer of research ideas at the university into viable commercial products. Peter Cowin and Peter Olive were the first to use this facility. Their ideas, though, needed to be piloted before embarking on any large commercial scheme. So in 1985 the NTC funded facilities in the university's zoology department. This was on a basis that the university would get a royalty percentage on sales, if and when the company was viable.

The department converted rooms to house the tanks for the production of the rag worms, and at the same time Peter Cowin obtained a Levenhulme Trust Fellowship to continue his research. They now started to look for sites that would be suitable for the establishment of the company. The university had a recirculatory system for the water needed for the tanks. Similarly, it was felt that in the commercial system there was a need for the provision of a constant supply of warm water. They came up with the idea of using surplus effluent water from power stations. This water was constantly produced as part of the power station system of cooling in the production of electricity. So Peter contacted the Central Electricity Board and the Scottish Electricity Board about possible sites. Quite a number of sites were indicated in Scotland and various parts of England. To show their serious intent, with the help of NTC, a business plan was produced for each site. They also now needed to get further advice on the business side of commercializing their idea. They talked to the Tyne and Wear Enterprise Trust and with their help drew up a cash-flow projection. Because there were possible sites in Northumberland, help was also forthcoming from the Northumberland Enterprise Trust.

They were now at the birth of the company and needed to decide on solicitors and accountants, the choice of which, it was felt, depended on how large they predicted their company would become. As accountants they chose Price Waterhouse, which was and is a large and prestigious international organization. They were fortunate in that both accountants and solicitors had confidence in them and were thus willing to defer costs until the company was set up and the cash flow had begun.

In 1986, still reviewing sites, they contacted the director of the Alcan Power Station at Lynemouth, Northumberland. The director, Anton Brooks, was looking to use the power station facilities to a greater extent and was extremely willing to take part in this new and unique venture. This was really the start of a very friendly cooperative partnership, the establishment of the commercial being of the company, and the coming together of all of the research and planning. A new business plan was prepared and this was entered in the "Innovation for Business Award" run by the Student Industrial Society and sponsored by Tate and Lyle and Touche-Ross. It won first prize. The benefit of the award was national newspaper and local television coverage, which increased the prestige of the company. Much of the business community was skeptical on viewing the business plan. Many soon realized the potential of the company and could see that there was the possibility of earning greater profits from rag-worm farming than from similar ventures with salmon. The business plan was also entered for the "Small Business Competition" run by *SHE* magazine and Lloyds Bank, and it won third prize. Another benefit of the awards was that the cash from the prizes contributed to the costs of setting up the company at a difficult and crucial stage. Their credibility was gaining acceptance and, with an introduction from Price Waterhouse, they approached the Department of Trade and Industry (DTI) for further grants. They were again successful in their bid. A major strength of the plan was the market research. Peter Cowin carried out a market survey, advised by NTC using questionnaires and direct telephone inquiries to dealers. They covered about 20% of the dealers and estimated that the rag-worm market had a retail value of about £2 million per year. It seemed that previously no one had carried out research on the market for rag worms in the U.K., although other countries had detailed data. They found that the estimate of the market in the U.K. was twice that which had originally been assumed.

Anton Brooks was quoted as saying "they were suitable cases for psycho-analysis." Clearly though, Anton could perceive the benefits from such a partnership in public relations for Alcan, using land that was redundant and providing new jobs in the region. So a formula was devised for the rent of land (at a very reasonable and minimal cost) and payment for the water. There were a few problems in the setting-up phase so Alcan waved the charges on the water to help with this phase of development of the company. The actual area for the site was covered with tall grass and was used as a dump for scrap metal. Alcan again helped by bulldozing and clearing the site. The first water tanks were installed in January 1987 and the first worms were sold in August 1987.

All, though, was not a smooth passage from then onwards. After a frantic selling of worms in August 1987 with initial good sales, the following two months had zero sales. At this stage, due to unforeseen conditions they lost 80% of their stock. Fortunately, because of their link with the university, this conditional problem was solved. They then hit a further problem, which drastically affected the stock, in that the water from the power station was too hot. There was also a drainage problem, in that the two sump pumps in the system were prone to failure. The staff of Alcan came to the rescue yet again. They found a way of providing cooler water and provided spare pumps to keep the drainage system working. Peter and his colleagues also managed to redesign the drainage system to make it more efficient. In the second year of operation, another disease was found in the rag worms. This again showed how vital it was to have retained the link with the university, where the problem was solved once more.

At this stage, because of the problems incurred, the company was in debt. So the partners had to request more money from the shareholders. They survived the financial crisis and continued to work with the university to suppress disease and increase productivity. The feeling was that total elimination of diseases would be hard to achieve. What they had to achieve, though, was management techniques to avoid catastrophic consequences for the business when there was a loss in stock due to disease.

The farm was progressing and in 1988 they applied to the DTI under the "Wealth from the Oceans" enterprise initiative to carry out a three-year joint project with Cell System Cambridge (a spin-off from the University of Cambridge), and a Norwegian-owned company, Golden Sea Products of Scotland. The theme of the project was the use of rag worms in agricultural foodstuff. The objective was to devise new products and increase breeding time.

In 1990 they entered the story of the company into the British Broadcasting Company "Radio Times and Radio Four Enterprise 90" competition in the section for small businesses two to five years old. They won this category with a prize of £10,000—which provided a great boost to their prestige and cash flow position. This provided further publicity both for the company and for Alcan. The company was proving good tenants. With environmental issues so prominent in people's thoughts, Seabait was to be, as Anton Brooks had so wisely contemplated, of increased value for Alcan in the field of public relations.

The company continued to grow. In 1993, with support again from the DTI, it expanded the site by 50%. Sales continue to increase, and although the company has not yet reached the figures proposed in the business plan, it has had the benefit of a higher price than expected. The partners are now increasing their effort to market the product abroad, with 5% of sales exported at the present time. The company has sent samples throughout Europe and Japan and has generated a great deal of interest.

It is interesting to look at the market competitors. Other similar companies have been set up in the U.K. and in the Netherlands. Seabait's mission, though, was to be better in business and research with continuous development using the links with Newcastle University. The competitor in the U.K. went into receivership with debts of millions of pounds sterling, and of the three companies in Europe only one has survived. Peter Cowin believes the success of Seabait can be attributed to having researched the product intensively. That the company can grow the rag worms more reliably and consistently better than anyone else in the world gives it confidence to do even better. He feels the partners were in the right place at the right time and crucially know exactly what they are doing. At the time of the formation of the company, there were both growing concern and media debate about digging the shores for bait and a market shortage of rag worms. Although Seabait rag worms are more expensive than those provided from other means, the Seabait product is of a very high quality. The mass market is for cheap bait, and Seabait is now close to competing with a low-price variety in some of its products. At the present time, the company can sell all they produce to the high-quality end of the market.

Seabait has consolidated its position, and, with the loss of most of its competitors, the partners see golden opportunities for at least the next five years. They also envisage future expansion on two further fronts—by licensing, which has already created interest and possible contracts, and by being a source for juvenile bait for other farms, a kind of franchising.

The objective, therefore is to develop the company's present site but to expand at a faster rate, to license their expertise and know-how, and hopefully to produce their bait more cheaply and sell at a lower price. The future for the company will hopefully continue to be very optimistic.

Within this novel example you can see the problems inherent in trying to form a company using an innovative idea. It is relatively easy to see that there are many tasks in the innovation process and that to break the process into invention, innovation, design, and diffusion is too simple. These components must be subdivided into categories. The idea of the production of rag worms was an original idea, but to implement the idea it was necessary to carry out market research, produce a business plan, find financing, and outperform the competitors. So in Chapter 2 we will provide a much wider breakdown of the innovation process. Fifteen tasks will encompass all of the various steps required to go from an idea through to a commercial product or service. We will then go on to discuss in detail these tasks in the rest of the book.

The case study reveals the importance of maintaining links. Seabait benefited significantly from its continued link with the university, which got the company through many difficult problems. In addition, the company-to-company link with Alcan again helped Seabait in many ways with the gain for Alcan being good public relations. A final point is that once success was achieved, the company could then expand the business through licensing and franchising. These are factors that will be in the innovation and technology transfer process defined in Chapter 2 and discussed in detail in subsequent chapters in the book

1.8 CONCLUSION

Technology is widely regarded as a key factor contributing to economic growth and international competitiveness. It is also accepted that for this growth to occur, research must be exploited to produce new products and services. Technology transfer becomes more and more important as the speed of all developments increases due to worldwide competition and much shorter product-to-market cycle times. However, scientific push is not the engine for technology transfer. Scientific discoveries may predate their uses by many decades. Yet, without scientific curiosity there would be no new technology to transfer. As long as there is scientific curiosity, technology will march on. This inexorable march of technology means that sooner or later products or processes will be overtaken and they will be out of date. Thus, the march of technology must be carefully observed so that new products and processes can be developed to take advantage of the changes and prevent the changes from taking advantage of the company. Taking advantage of the changes requires that a positive attitude be taken towards the process of innovation and technology transfer.

The extent that innovation and technology transfer will contribute to the creation of wealth depends on the way countries are able to develop, select, acquire, and disseminate know-how through their national systems. It is also dependent on other such factors as management attitude, research infrastructure, and financial systems. The whole process that leads to the successful exploitation of ideas is very complex, as we have begun to show in this chapter. In the rest of the book, we will deal with and discuss all of the relevant factors that make up this interwoven web, which can be called the structure for innovation and technology transfer.

References

[1]   Colgrave, B., and R. A. B. Mynors (eds.), *Ecclesiastical History of England*, Oxford, England: Oxford University Press, 1994.

[2]   *Business Week*, August 3, 1992.

[3]   Arthur D. Little, Inc., *Management Perspectives in Innovation*, Cambridge, MA, 1985

[4]   Morita, A., E. M. Reingold, and M. Shimomura, *Made in Japan*, New York: Dutton/New American Library, 1986.

[5]   *The 1993 UK Scoreboard*, Company Reporting, Ltd., Edinburgh, June 1993.

[6]   Braun, E., and S. MacDonald, *Revolution in Miniature: The History and Impact of Semiconductor Electronics*, 2nd Ed., Cambridge, England: Cambridge University Press, 1982.

Selected Bibliography

Davis, W., *The Innovators*, London: Ebury Press, 1987.

Henry, J., and D. Walker (eds.), *Managing Innovation*, London: Sage Publications, 1991.

Tushman, M. L., and W. L. Moore (eds.), *Readings in the Management of Innovation*, New York: Ballinger, 1988.

Chapter 2

Innovation and the Technology-Transfer Process

2.1 INTRODUCTION

When IBM grants a license to the government of Taiwan, the computer corporation undertakes a transfer of technology. When a French cheesemaker passes on recipes to Japanese firms, a transfer of know-how is performed. When Boston University organizes continuing-education programs in Belgium or Russia, the school undertakes a form of cooperation of technical and educational assistance. The flows of technology and know-how are various and wide ranging. Technology transfer is a focal point in the management of technology and innovation in companies. Innovation has been shown to be a major priority for the most successful organizations. This is true whether they are in the U.S., the Far East, or Europe. It is estimated that between 30% and 40% of the turnover of the best performing companies is generated by products that have been launched in the previous five years. So rapid development through the acquisition and integration of new technologies will increasingly be a key to the success of an organization.

In this chapter we will provide an explanation of the innovation and technology-transfer process and an overview of how this process will be discussed in the rest of the book. There are many definitions of innovation and technology transfer and lots of ways that people try to show how the process can be broken into various categories. These breakdowns are produced in an attempt to help industrialists understand the complex issue of innovation and technology transfer and the ways that they may translate the process to the operations of their companies. Confusion does arise between innovation and technology transfer because the two titles are often interchanged out of context. We would subscribe to the view that innovation is the whole process from idea to finished product while technology transfer is the means of achieving innovation. In other words, technology transfer is a subset of the process of innovation as we show in Section 2.2. In our experience, the innovation process described in this section is the one that is easiest to understand and is for many companies the most natural.

2.2 THE INNOVATION AND TECHNOLOGY-TRANSFER PROCESS

To have the greatest possibility of success in the adoption of new technology, the management team must look at all the options open to it. A full understanding of the innovation process provides the tools to help them along this road. For any organization, a very useful checklist is provided in Table 2.1.

In some examples, the process carried out has followed the breakdown chronologically as a sequence of step-by-step events. This, though, is not typical—events will overlap in time and could be implemented in differing orders. Therefore, the numbers in the list are for reference purposes only. The areas of responsibility are in reality somewhat hazy. There can be many different people and organizations involved, and there can be overlaps in responsibility for some of the tasks. Thus, the ones given are for example and are not intended as any definite rule.

The innovation process is quite long and can be very complex because the individual processes do not necessarily follow each other in chronological or logical order in practice. It can, though, be split into three definitive areas:

- Points 1–4, external factors to the company—the options and opportunities. In company practice, this phase could be phrased as "where we are going."
- Points 5–9, internal factors—the feasibility of the options (i.e., "Can, and how, do we do it?").
- Points 10–15, implementation—the launch of the idea (i.e., "out into the marketplace").

The list contains business, as well as technical, processes. This makes it the innovation process rather than one purely for technology transfer. A great deal has been written on business issues such as marketing opportunities and the management of change. Little attention has been paid to describing the technical aspects of change (i.e., technology transfer). In this book we will concentrate far more on technology transfer, which is characterized in the list by points 2, 3, 4, 8, 9, 11, and 13. Our premise is that technology transfer is a subset of the innovation process, namely:

- Identify appropriate technology;
- Evaluate technology;
- Secure technology;

Table 2.1
Innovation Process Checklist

Task	Responsibility
(1) Recognize a market need/opportunity	Company initiative
(2) Identify appropriate technology	Project manager or consultant
(3) Evaluate technology	Manager/technical director
(4) Secure technology inhouse R&D joint ventures agency/distribution inward licensing	Agency, consultancies, higher education institutes (HEIs)
(5) Create business plan	Managers, accountants, external financiers
(6) Obtain finance	Banks/investment organizations, government agencies
(7) Establish market potential, demand, growth, competition, and trends	Market research consultants, project consultants, marketing/sales department
(8) Protect technology confidential disclosure exclusivity patents/trademarks	Patent agency, legal advisors
(9) Produce prototype, product development	Innovation center, design councils, R&D units
(10) Establish management systems	Managers, investor/sponsor, consultant
(11) Obtain technology-awareness training	HEI's CAD/CAM centers, consultants
(12) Obtain business training	Business schools, consultants
(13) Organize product-specific training	HEIs, specialist training organizations
(14) Launch product, process, or service	Company
(15) Monitor and control total operation	Project consultant or manager

- Protect technology;
- Produce prototype and develop product;
- Obtain technology-awareness training;
- Organize product-specific training.

However, the technical aspects cannot be completely divorced from the business ones, which will be concisely discussed though in less detail.

A very good example of the process is given in the case study of Seabait in Chapter 1. If you check back to this case, you will see that most of the tasks in the process were at some stage encountered in the development of Seabait. It was important for the company that the technology was transferred from research at the university. It was equally important for the success of the company that the partners carried out the business, organizational, and strategic tasks. To recap from Chapter 1, we stated that the farming of rag worms was an original idea, but implementing it required carrying out market research, producing a business plan, finding finance, and being better than the competitors. Important from the aspect of technology transfer was not only the transfer of the idea but the continued link with the university, which got the company through difficult problems. Also important was the company-to-company link with Alcan, which again helped Seabait in many ways and Alcan in the good public relations generated. Once success was achieved, Seabait could then expand the business through licensing and franchising. Thus, the act of development from idea to product could best be described by categorizing the various phases into the tasks given in Table 2.1.

The major points in technology transfer are searching for, finding, and then evaluating the new technology (steps 2, 3, and 4). These are discussed in detail in Chapters 3–6. Any new venture, though, must be associated with a business plan and a method of financing, which are discussed in Chapters 6 and 7. A fundamental act with a new product is to protect it by law using patents (step 8). There are many other facets of protection, such as copyright and having employees agree not to sell secrets or use any part of a new design from a previous employment in a new job. These important issues are discussed in Chapter 9. A new product may be so innovative to a company that it needs to have a prototype (step 9) and the company will have to assess how it will develop the new service or product. To be successful, it must also ensure that the risks of failure are minimized. These ideas are discussed in Chapters 11 and 12. Imbedded within the process but often forgotten is the human element. The need to convince management and workforce that change is good is dealt with in Chapter 12. The absolute necessity to train employees in different practices for the new product or service (steps 11 and 13) is discussed in Chapter 14. The final, generally emotional and momentous occasion, such as the rolling out of a new aircraft type, is to launch the new product or service. To get to this stage requires good project management. These two issues are dealt with in Chapter 13. With all projects there has to be a form of control for costs, quality, and development time. Some of the more traditional ways of control do not function well in this new and fast-moving industrial and commercial world. So, finally, in Chapter 14 we look at control and performance measures, with particular emphasis on methods for innovative companies.

At this stage, rather than rush into further theoretical aspects, let us look at some examples that should help to enhance the understanding of the process. The examples illustrate in particular project-appraisal situations, which have implications for the acquisition of technology. They relate to strategy decisions, which need to be made prior to technology-transfer activity.

2.3 EXAMPLES

2.3.1 Example A

In this example, you will see a very interesting point—as well as showing some of the processes, this example illustrates how timing or luck can play a major part in the fortunes of companies in the transfer of technology. The company had initiated the search for new technologies at the very same time the inventor-supplier of a prototype product, which would prove ideal for the company, was searching for a company to mass manufacture and distribute his ideas. The eventual realization of this innovation was the result of the meeting and combination of two separate organizations—each of which was, in a sense, looking for the other.

Alphus Instruments, had achieved market leadership in high-quality inspection equipment. Its position in the market, though, had begun to decline over the previous ten years. It was eventually taken over by a larger manufacturer in 1992. At that time the firm's product range was technologically outdated and the market segments in which it concentrated were declining, at a rate of about 20% a year. The new managing director, after consultation with group headquarters, decided to undertake a modernization program for the entire product range. Staff members were asked to study technologies that had the potential to form the basis of a new product range and to research possible new (to the firm) markets which the company could enter.

The external search, using information systems such as the patent database (developed further in Chapter 4), produced a map of the metrology market (the company's market) and a number of potential technological developments in various stages of completion. The decision about which development the company would pursue was mainly technology driven in that the new products, a range of advanced measurement devices, incorporated a combination of developments in computing, electronics, optics, and laser technology (all of which were new to the company). However, they were also targeted at the same metrology market as for the company's existing mechanical products; the company's eventual aim was to replace them.

The invention of the basic device and the technological development of the prototype had been made by a university lecturer in engineering via his own company, which was formed to develop products from his research. The lecturer had been discovered during a technological search initiated by the managing director.

Alphus Instruments had very limited capabilities in that it had no skilled staff, no manufacturing equipment, and no distribution or sales expertise in the new technologies. So considerable management time was devoted to the critical consideration and selection of the means of acquiring the required technologies, as well as the method and source of manufacturing.

The implementation of the chosen innovation was complex for the management of the company, involving relations with group management; negotiations with the inventor, suppliers, and legal representatives; technical development and, more critically, control of the manufacturing process; and managing the organizational and staffing implications (particularly training) for the adoption of the new products.

So, in reflection, what has this example shown, in comparison with the model for the technology-transfer process? In the external phase, we see the company looking for a new opportunity, searching for the technology, and linking with a new technology provider (who was also looking for a partner). In the internal phase, the company had to decide on which products to develop, look at a prototype, protect the technology, and establish ways to develop the product. Then, in the implementation phase, the company had to establish the management systems, train the workforce, control the manufacture, and launch the product.

2.3.2 Example B

Here, one of the world's largest electronic companies uses a one-page format to document and plan an innovation project. You will see that the organization has identified a new area from which it hopes to benefit commercially.

Objective of Project:
Develop robust electric pressure sensors

Start Date: January 1, 1992
Finish Date: September 1993

Budget (Manpower and Other Costs):
8.5 person years plus £400,000 equipment.

Nature of Project:
To develop a commercial product based on *our* new sensor technology.
The project will develop the sensor element to commercial production

level and then develop a product design with local integrated circuit (IC) based calibration and data error correction.

Key Stages:
Prototype product: May 1, 1992
Pilot production of sensor element: December 31, 1992
IC design: December 31, 1992
Preproduction prototype: June 30, 1993
Pilot production: September 30, 1993

Expected Benefit:
Sales of £1.2 million in 1994, rising to £5 million a year by the end of 1995. Knock-on benefits in sales of data loggers.

Linkages to Other Projects:
IC design is dependent on acquisition of new CAD system.

Principal Hurdles and Risks:
Yield of sensor production process
Shock resistance of sensor elements
Cost of product assembly

By the looks of it, the company doesn't have to search for the new technology (consider the first sentence under "Nature of Project"); this will be provided in-house. The managers have obviously identified a market need because they state the expected profits from the venture. They have noted what will be needed to produce the product, namely the new CAD system. A crucial aspect of the plan is the key milestones—note the production of a prototype and the pilot development, which inevitably are vital ingredients in the future success of the project. Finally, they have estimated the costs and the possible problems of the project.

This gives a good outline of the proposal but more detailed financial, marketing, and technical data would need to be provided in a business plan to ascertain fully the consequences of this new product for the organization. The implementation side is very much neglected in the simplistic plan but would obviously need addressing as part of the overall plan and company strategy.

How can innovative ideas be generated for a company to make technical changes? Example C illustrates how a major European company generated and evaluated ideas.

2.3.3 Example C

This company was experiencing a shortage of new product proposals. It decided to adopt a program to overcome this problem. It consisted of five stages:

1. *Concept generation.* This involved over 200 employees using a variety of approaches, such as brainstorming and Delphi methods (see the Selected Bibliography for books on this topic), to generate ideas. Over 600 were produced.

2. *Idea screening.* A preliminary screening of the ideas was undertaken using three criteria: feasibility, familiarity, worthiness of consideration. This reduced the number of ideas to about 100.

3. *Relevance to the company.* The remaining ideas were categorized in relation to how well they fitted the corporate strategy as it existed or as it might be in the future. It was accepted that some of the ideas might be of sufficient merit to warrant a modification to the strategy, but in such cases they would need to offer exceptional benefits.

4. *Evaluation.* The ideas were then submitted to the company's formal evaluation procedures, and those satisfying them were allocated priorities.

5. *Project selection.* A few proposals were then selected for development in relation to their match with the R&D and market portfolios and the availability of resources.

In this example, the process may appear expensive and wasteful, as only a few of the 600 original ideas were developed into commercial products. However, it did result in products that would otherwise not have been developed, and it was a huge public relations success with the company's employees because it generated high morale and a feeling of partnership among them.

The technology process has many facets. Where possible, we aim provide good practice ideas, provoke thinking, and even illustrate where ideas have failed and why.

To recap the process: We feel innovation is the whole process from an original idea through to a product on the market. This will include many tasks that involve management, organization, business and strategic factors. The means of achieving innovation is through technology transfer. The tasks here will be to find the technology, protect it, and then develop it to form a new product or service for the marketplace.

It must be stated, though, that innovation by its very nature has an element of chaos, or nonlinearity. No two innovations seem to follow the same path from conceptualization to commercialism, even within the same organization. Therefore, it is unlikely that technological innovation can be entirely planned whatever strategy is followed. It is like a jigsaw puzzle where all the pieces are known, but there is no unique way to put it together. In this book we will give you the pieces, which we hope will give you the opportunity to complete the puzzle in the way that suits you best.

Most companies need to establish the following: What are the new opportunities or options? Do we need to change ? Have we got a new product that is going to make us a profit? And if so, how are we going to get it?.

2.4 COMMON BARRIERS TO INNOVATION

If we accept the premise that technology transfer is necessary for companies to be innovative, remain competitive, and gain competitive advantage, what are the factors that block innovation? Opinion varies on how many factors pertain to this and how to categorize these factors. We believe there are five major barriers:

- Management attitude;
- R&D effectiveness;
- Short-term pressures;
- Resistance to change;
- Poor information flows;
- Weak links with customers and suppliers.

Management attitude is crucial to technology transfer. Unless management is willing to accept that products and process must change, the company is virtually doomed. It may take some time, but other producers will inevitable take advantage of opportunities ignored by a neophobic management. For example, consider the demise of the British motorcycle industry and the ups and downs and again ups of Harley-Davison in the similar industry in the U.S.

Even a management that does embrace change may have weaknesses. One of these weaknesses may be towards new products that have been researched and developed inhouse. However long a product takes to bring to the marketplace and however many resources are invested in it, as soon as the product reaches the market, all of the research and the solutions to technical problems are known to the competitors. The opposition thus knows how you have tackled a particular problem and can take advantage of this in their own solution. As a result, the competitor can bring out a similar product in a shorter timescale with far less R&D costs. We have seen many examples of this in the consumer electronics and goods packaging industries. Therefore management must not become so attached to a new product that it ignores the continuing march of technology. Management also cannot afford to be complacent. A company that relies on a product that has done very well for the past four years may find that the product is obsolete in its fifth year. The company must continuously look forward.

Often R&D is regarded as an activity that only needs to be undertaken every five years or so, or when a new product is seen to be needed. It can be expensive, but it needs to be continuous to be effective. Companies that take a short-term view may not be willing to spend on R&D when the payoff appears to be a long time away. This short-term view means that many companies are reluctant to innovate and gamble on new products—a position that leads to noncompetitiveness and bankruptcy.

Information is the lifeblood of technology transfer. The better the information flows, the more likely appropriate and timely changes are to be made. If the information flows are inadequate or the response to suggestions is negative, technology transfer is likely to be hindered. It is essential that the opportunities for information to move around a company are maximized. Similarly, it is also essential that the opportunities for information to come into the company are maximized. It is useful, therefore, for a company to keep its formal and informal information networks under constant review. Marketing is a form of information flow from the company outwards. A good product can be destroyed by poor marketing. The Sinclair C5 invented by Sir Clive Sinclair in the U.K. was an innovative product but it was not a car, which is what it was marketed as. The C5 was a buggy-type vehicle, which would have been useful in warehouses, shopping precincts, and airports, where some are still used. A product designed and marketed for one purpose that turns out to be used in quite a different way, or for quite a different purpose is known as *reinvention* or *secondary invention*.

The period of the late 1980s and early 1990s could be called "the years of quality and customer service." A great majority of organizations realized that profitability was dependent on good customer and supplier links. It is still the case that no matter how innovative and excellent a new product might be, it will unlikely be a success if the company does not provide a quality service and good links and communication flows with its customers and suppliers. There is the need for high quality in product, customer and supplier links, aftercare, and maintenance if a company wishes to be successful, which generally means profitable. To begin with, though, someone has to recognize an opportunity.

In small to medium-sized enterprises (SMEs) there are additional reasons given for not using technology transfer, in descending order of those most commonly cited:

- No time;
- Too expensive;
- Current products meet needs;
- Unsure how to achieve transfer;
- Too much red tape.

Many managers in SMEs feel they have insufficient time to meet daily requirements and therefore find it difficult to devote time to long-term thinking. This tends to be a major reason why business growth is not sufficient in many SMEs, leading to financial problems.

2.5 RECOGNIZING A NEED OR OPPORTUNITY

Before going into recognizing a need or opportunity—*technology scanning*—let's refresh our minds on the broad reasons for innovation and why we need to look at options. In Chapter 1 we listed the reasons under six headings:

- Scientific changes (either major ones or internal research);
- Competition;
- The market;
- Legislation;
- Human inquisitiveness;
- Innovation as a company policy.

Research literature on innovation often concentrates on the balance of influence between technology push and market pull (i.e., the need to change because of new technologies versus the need to change because of new markets or new ideas and influences in the marketplace). The dominance of either of these modes is likely to reflect the culture of the company and the relative strengths of internal sections. A mutual understanding and a close relationship between the two influences is essential to ensure *both* technical innovation *and* customer-led changes. Either way a company recognizes a need to change and is scanning for a new product, service, production method, or market.

From the technology push side we have *monitoring information about external research and developments*. This is often the responsibility of a person or section with a title encompassing R&D. However the role of the individual throughout the organization is crucial. An objective of any organization should be to establish a supportive environment where staff can keep their knowledge up to date and establish external linkages. This can be difficult to justify to some senior management in times of economic stress, but despite the nonspecific nature of the benefits, such activities are as essential as market research or close relationships with customers.

Examination of the information-flow processes within companies usually reveals that there are some individuals who, because of their natural interest and curiosity, read widely and have an extensive network of informal professional contacts. As a consequence, they acquire a great deal of random information which may have little relationship to their formal duties. However, when a problem arises they are often consulted by colleagues. On some occasions they may be able to provide an answer, but more frequently they can suggest an external contact who is able to assist. These people are called *gatekeepers*.

Gatekeepers cannot be appointed but their importance needs to be recognized and their use of company time not discouraged. More broadly, an encouragement of a controlled gatekeeping role for all staff can greatly assist upward communication flow.

A good example of the importance of gatekeepers (and a warning about the need to have gatekeepers throughout the organization) comes from an international company based in the north of England. The company with leading-edge products in polyol and waterbased coatings [1] had an enviable record of investment. However, its main innovator and gatekeeper died. This caused enormous problems in the development of new products and even problems in the continuous development of their present products. Therefore, the company had to seek out and try to use external consultancy to continue development. Fortunately, in this case, this was successful but clearly the opposite could easily happen and a company's future could be put in jeopardy.

International statistics regarding *inhouse R&D* certainly paint a bleak picture for European companies generally, whether the statistics are based on patents registered, research expenditure, or peer assessment of product success.

The link between research, invention, innovation, and competitiveness is, of course, complex. However, in support of well-planned research, the 1992 Department of Trade and Industry report on innovation in Britain [2] ended with the following statement:

> The study has confirmed that the concept of innovation dominated by 'technology push' has tended to shift towards one where there is a more even balance between this and 'market pull'. However, it has also shown that technological advantage can still be a key factor for innovative success and its value should not be underestimated.
>
> A number of the project's best practice companies rely on technology (existing and new) to generate a stream of new ideas. Some of the less successful companies could improve their position significantly by upgrading their technological competence, by forging links with other companies and academia, and with help of Government support schemes.
>
> In the technological field, the study has confirmed the importance of accessing the global technological base for new ideas.
>
> The value of technological advantage, however, must be accompanied by an organizational structure which can best exploit it.

This could be stated for most countries in any part of the world. Even an active research effort within a company will generally require further technology from outside the company. Other chapters of the book will consider the various routes by which this technology transfer can take place.

From the market-pull influence side, we must consider *recognizing market need and competition*. This does not necessarily lead to the eventual use of new technology, but will usually require the appraisal of some technology-based options.

The importance of competitors is a constant theme throughout the rest of the book. Much of this importance stems from the use of competitors to provide the best practice comparisons for benchmarking all company activities (which will be discussed in Chapter 3), but the monitoring of their activities and products can just as importantly reveal trends in the market.

The innovative company envisaged by this book as a whole is both customer-led and responsive to change. Thus, an awareness of changes in the market should be more easily achieved in such a company than it would be in a noninnovative company. General techniques, loosely grouped as environmental scanning, are used to keep a regular watch on the market and spot potential areas of market need.

The word environmental has nothing to do with ecology or political green parties. It means the business environment in which the company operates. The wide range of information found in scanning the market environment will also be useful in compiling parts of the business plan for the new development.

Table 2.2, adapted from a major U.S. survey of successful small companies, shows the type and frequency of information collected by these companies.

Table 2.2
Information Collection

Type of Information	Never	Irregular	Periodic	Routine	Continuous
Remote market place information trends					
Demographic	50%	31%	15%	0%	4%
Population	65%	23%	12%	0%	0%
Sociocultural	54%	34%	4%	4%	4%
Immediate market place information					
Customer characteristics	12%	16%	32%	12%	28%
Customer opinions/attitudes	11%	11%	22%	34%	22%
Customer behaviors	14%	14%	26%	26%	18%
Market growth	7%	11%	45%	26%	11%
Market segments	12%	17%	54%	17%	0%
Competitor's strategy	15%	4%	12%	50%	19%
Competitor's sales and market shares	12%	20%	12%	52%	4%
Competitor's products	0%	8%	13%	33%	46%
New entrants	12%	12%	24%	36%	16%

After: [3].

The favorite sources of information are listed in Tables 2.3 and 2.4. (Note that the totals in the two tables below add to more than 100%. This is because multiple responses were allowed).

Table 2.3
Most Popular Sources for Gathering Market Information

Source	Response
Customers	89%
Business contacts	82%
Competitors	82%
Suppliers	74%
Employees	70%
Trade representatives	67%
Business affiliates	63%
Industry experts	60%
Manufacturer's salesforce	59%
Social networks	48%
Friends	48%
Subcontractors	41%
Consultants	33%
Family members	30%

After: [3].

Table 2.4
Most Popular Impersonal Sources for Gathering Information

Source	Response
Trade magazines	96%
Sales brochure	70%
Advertising	63%
National newspapers	60%
General magazines	56%
Journals	56%
Local newspapers	52%
Manufacturer materials	52%

Table 2.4 (continued)

Source	Response
Catalogs	51%
Annual reports	41%
Government publications	22%

After: [3].

We can deduce certain things from the information in the tables and from our own experiences. That a range of many different sources are used by companies is obvious. Formal methods of information gathering are rare. This appears true even in successful companies. The reasons are predominantly lack of awareness and cost. Companies generally wish that there was more formal information available on broad trends like population or social characteristics. There is a major question of whether many companies have the expertise to interpret such data when it is available. In any case, customer needs, competitors' products, and market growth are seen as the most important factors to scrutinize. Certainly among SMEs, the main identification of new product development used by about one in two companies is customer feedback with sales team feedback a long way behind.

If innovation is so crucial to the economy of a country, as many politicians claim, what really are the benefits to companies that are innovative?

2.6   VALUE TO THE COMPANY

The main benefits lie in:

- Increased competitive advantage;
- Improvement in quality;
- Cost savings;
- More flexibility;
- Better service to customers;
- Reduction in lead times.

Companies that make strong investments in innovation do in general achieve better cashflows and earnings performances, and, above all, greater long-term growth than companies that do not. In the U.S. it has been established that investment in R&D correlates well with long-term growth. The reverse has also been proven. Costs saved by cutting R&D can improve profitability in the short term but tend to have an impact on future products and, thus, future profitability. Growth should be the key aim because growth is what leads to a company dominating or having a major influence on the world market. The impact that companies from Japan have had on the world market are due in part to their determination to control markets at the expense of lower profits. In times of recession, the art of laying the foundation for future success is the ability to keep investment in R&D. Often in economic downturn, new product development programs seem difficult to justify and are put in storage or scrapped, but the brave companies that do still try to be innovative are generally the huge success stories of tomorrow. It is not just in manufacturing that such investment can be fruitful. It can also be the case in service industries. In the U.K. the department store Marks & Spencer has been very successful with the innovative use of chilled food retailing and in the U.S. Federal Express has made huge improvements in customer service with investments in package tracking and sorting.

Let us look at one company's ideas on how it sees the value of investment to future profitability. They decided the way forward was to invest more of its turnover in R&D than its competitors were, invest in a new product, and use technology transfer to bring in a new computer-based ordering system.

2.6.1 Example D

Core Construction is a company specializing in the sale of building products. The company decided to spend 6% of its turnover on R&D, compared with 3% generally allocated by its competitors. The company is investing heavily in a major new product that it hopes will increase sales by 40% in three years. It has also searched for new technology to provide customers with an online ordering system using remote computer terminals. This will allow it to compete with companies that, at the moment, have a larger volume of production. The aim of the company is to become the country leader in specialty commercial building products. To do this, the managers believe they need to launch a continuous stream of new products and invest in new process machinery and a new plant to ensure production flexibility. They have also anticipated new environmental legal requirements by investing now in the development of cleaner technologies, minimizing waste, and exploring recycling opportunities.

The results so far with this kind of thinking have been good. Nearly 70% of the current product range is less than three years old. Sales in existing markets rose 20% last year, despite a static market and the threat of increased competition.

The success of this company is a direct result of innovation and technology transfer. If the company doesn't innovate, its products will become mere commodities and the company will have to compete solely on price. The profitability and growth prospects of the company would be seriously threatened.

To provide more evidence of the benefit of innovation, let us concentrate on the aspect of our main theme, technology transfer, and its value to companies. We will do this by considering the advantages gained through technology-transfer schemes pertaining to the introduction of flexible manufacturing systems (FMS) to companies. By introducing FMS, substantial savings of an average of 65% were made in throughput (factory door to door) time and an average of 50% in lead time to customers. There was a smoother flow of material through the factory with less buildup of material waiting for machining. Along with less setup time, this smoother flow allowed an inventory cost savings of 70%. Utilization of machinery was much higher, so machines were stopped less for resetting and the number of machines standing idle was reduced considerably. Most importantly, there was a major improvement in quality. FMS-related sales figures increased. In fact, the FMS allowed firms to diversify their product ranges and move into entirely new markets and products. Companies were able to calculate that, through the introduction of FMS, unit costs were noticeably reduced. The introduction of this new technology did not cut jobs as many feared, rather the cost savings it allowed enabled companies to expand and actually take on staff. In addition to all of this was a bonus for the long-term stability of the company, in that future-development capability was improved, allowing growth to continue through more rapid product-innovation cycles and better prototyping facilities.

Technology transfer is at the heart of innovation and innovative companies are the ones that now survive and become established for many years to come. Clearly there are financial gains to be made by the company through the introduction of new technologies resulting in new products. The community also benefits with improved job prospects and thus increased prosperity.

2.7 CONCLUSION

In this chapter we have placed an emphasis on the need for companies to think ahead and be innovative. This is because business statistics have clearly illustrated a statement once made by Forsyte in the *Forsyte Saga* by John Galsworthy [4]: "If you don't think about the future, you won't have one."

We have introduced the innovation and technology-transfer process and showed that it is a very complex set of events.

In Chapter 3, we will begin to look at the first of the factors that form this valuable process of innovation: identifying and monitoring technology information. Information flows were described earlier as barriers to innovation, but they are also the thrust in technology transfer. If a company cannot find the technology it needs for a new product or service, it cannot develop or grow.

References

[1]   Miller, R. W., *Lubricants and Their Applications*, New York: McGraw-Hill, 1993.

[2]   *Innovation—The Best Practice*, report produced by the Department of Trade and Industry and the Confederation for British Industry, London, 1993.

[3]   Brush, C.G., "Marketplace Information Scanning Activities of New Manufacturing Ventures," *Journal of Small Business Management*, Vol. 30, No. 4, 1992, pp. 41–52.

[4]   Galsworthy, J., *The Forsyte Saga*, London: Oxford University Press, 1995.

Selected Bibliography

Research. Technology Management, Industrial Research Institute, Washington, DC.

International Journal of Technology Management, Inderscience Enterprises, Ltd., Geneva.

Delphi Method Used in Forecasting 1970–1988, Washington, DC, National Technical Information Service (ref PB87-868303).

Granger, C. W. J., *Forecasting in Business and Economics*, London: Academic Press, 1989.

Chapter 3

Identifying and Monitoring Information

3.1 INTRODUCTION

In Chapter 2, the most popular commercial and technological information sources of managers were described. In this chapter, we will be looking at information collection as a distinct strategy. This strategy will ensure:

- A well-organized approach for monitoring technological and scientific intelligence;
- Good research and development organization to encourage information flow (both of internal and external information);
- Accurate knowledge of the company's own technological position and capabilities (through technological reviews or audits);
- Efficiently organized information on markets and competitors.

As with many other aspects of technology transfer, there tend to be differences between countries on overall approaches to the organization of technological/scientific information. Differing ideas on the role of central government is a paramount reason.

For example, there has been little formal development of technological/scientific intelligence by the U.S. government for use in the private sector. In addition, as Herring [1] points out, by 1993 probably only about 3% of U.S. corporations had fully developed intelligence systems. This compares unfavorably with, say, Germany or Japan. Although American companies have continued to lead on product development and stay at the cutting edge of knowledge, this lack has caused them to look only at other American companies or labs for new technology information.

In Japan, intelligence units are much more common, with well-staffed *listening posts* in overseas offices. The units are usually administered from within the planning department. R&D centers have also been established overseas, often drawing on the experience of local researchers (who, of course, carry knowledge and ideas as ex-employees of local competitors). The U.S. firm Motorola has now established a major research center in Japan with similar advantages. Herring points to a 1991 survey in Germany revealing that half of the companies surveyed had intelligence programs. A third of these were centralized in specific departments. The respondents indicated that the average warning time about a new technological development in the nonchemical/pharmaceutical sector could increase from 17 to 33 months if an intelligence program existed.

An admirably comprehensive American example is Baxter International's corporate technology sourcing (CTS) department. This was set up in the early 1980s with six staff members and a half-million dollar budget to scan and monitor worldwide technology developments. They scan 5,000 new technologies a year (using relationships with research labs/universities, databases of published scientific/ technological papers, professional meetings, and coverage of trade conferences). Eventually perhaps between 7 and 10 ideas may be turned into products.

Regional technology advice centers or other similar government support agencies can advise on or carry out literature searching for technological information.

There has been debate about whether intercompany partnerships help or hinder technological intelligence in a company. While there are advantages when one partner brings in new technological expertise (in exchange, perhaps, for market access or production facilities), in other cases it might be that the ready access to this particular partner technology may cause the technology intelligence program to atrophy.

3.2 RESEARCH AND DEVELOPMENT

Both research and development are terms intimately linked to innovation and technology transfer. At its purest, *basic research* may not be tied to any immediate product development but will still be considered vital to: (1) ensure that knowledge on a subject is readily available, and (2) stimulate new product opportunities.

There may also be a need for *common elementary technology research* for a variety of products and for many production systems. Kurawaha et al. [2] describe how Hitachi organized its research in Japan. The structure was:

- Independent research:
  - Basic research plus common elementary technology research;
  - At the advanced research lab, the central research lab, and at the nine corporate labs;
  - Funded by head office within agreed long-range corporate-research strategy.
- Commissioned research:
  - Mostly new product development;

- At the central research lab and the corporate labs;
- Sponsored by the factory, subsidiary companies, etc.—but within context of research strategy.
- Product improvement:
  - In the development departments at the factory;
  - Funded from factory budgets but must refer to the context in the research strategy.

Basic research and common elementary technology research can require a tremendous financial commitment in many sectors. For example, Canon will spend about 11% of sales on R&D. The norm in Western engineering companies would be nearer to 3%. But before committing funds for any sort of research (internal or external), even a small company needs a research policy.

At the broadest level, the company may choose to be a technical leader, a product innovator, a nicher, a follower, or a combination of these.

Larger, distributed companies will need to choose whether to have:

- Centralized research labs, which allow coordination and concentration of resources on broad strategies but require the linking mechanisms (i.e., good information flow) with production centers, customers, and suppliers;
- Labs distributed to teams, which permits closer contacts but also raises the risk of duplication or poor coordination.

3.3 ROLES OF STAFF IN INFORMATION FLOW

In Chapter 2 one of the three major stereotype roles in information flow—the technical gatekeeper—was introduced. The two other major roles are the *godfather* and the *champion.*

The godfather is a senior manager who watches over and protects the technology-transfer project and informs or convinces other senior staff of the value of the new technology. The role of the godfather may continue throughout the whole transfer process or may be limited to its development phase.

The champion, who is usually at the middle-management level, is skilled and committed, and may have to take over from the godfather for the implementation phase. The volunteer champion is likely to bring the benefit of a knowledge of internal politics, in comparison with an appointed external champion. Good communication and interpersonal skills are required (not always a feature of a technology-/information-oriented staff). The gatekeeper would certainly have the required technological knowledge but may be reluctant to assume the management implications, even in a flatter team hierarchy.

3.4 TECHNOLOGY REVIEWS (OR AUDITS)

An important technique for assessing the current state of a company's technological health and for assessing future technological innovations is the technology review (or audit).

It typically has three parts:

(R1) Review of competitors' technology (what have they got?);
(R2) Review of company's own technology (what have we got?);
(R3) Review of state-of-the-art technology (what could we have?).

Later discussion in this chapter will link R1 with *benchmarking,* and the links of R3 will be discussed in Chapter 4.

After a review has been set up and clearly supported by senior company staff, the following eventual uses for the review should be emphasized:

1. Briefings for key need-to-know managers and professionals (this is what we've got, what they've got, and what we could have);
2. Briefings for suppliers that provide needed technology (can you get us what we could have?);
3. Briefings for R&D teams, including product-design staff (can we find what we could have?);
4. Briefings for external stakeholders, such as customers, financial institutions, and government agencies (can we afford to get/not get what we could have?);
5. Any inhouse technology-transfer activities, such as seminars and training programs (hey, look everybody, this is what we could have!);
6. Integration into evolving technology and business strategies (we're going to get what we could have and use it!).

The most effective transfer of technology comes about through people links; therefore, technology reviews keep people informed, empower technology awareness, and produce better decisions.

A basic tool of technology management, the technology portfolio, can be easily used to categorize each of the company's technologies according to its competitive position and business attractiveness. Usually, the technology is presented in a four-cell, two-by-two matrix, with business attractiveness or technological importance along one axis and the competitive position along the other (see Figure 3.1). An assessment can then be made as to which technology projects should be emphasized in company business strategies.

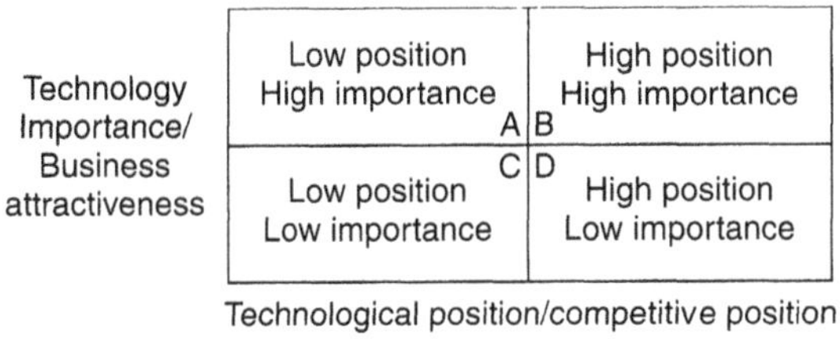

Technology Importance/ Business attractiveness	Low position High importance A	B High position High importance
	C Low position Low importance	D High position Low importance

Technological position/competitive position

Figure 3.1 The technology portfolio.

Technologies that fall into cell A are considered important and attractive as current projects and programs, yet the company is not currently in a strong competitive position with respect to them. Thus, R&D resources and emphasis should probably be increased to gain a more competitive edge.

Technologies in cell B are considered important and the company currently sustains a strong competitive position with respect to them. Current levels of resources are probably sufficient to sustain these technology projects, but the company must be prepared for possible competitive attacks.

Technologies in cell C are not considered important and the company does not maintain a strong competitive position with respect to them. They should be considered for elimination from the technology portfolio of the company.

Finally, those technologies in cell D are not considered important, yet the company maintains a strong competitive position with respect to them. Senior staff responsible for R&D and those responsible for strategic planning should consider the reasons for continuing these projects before either moving them to cell B or eliminating them.

You might consider briefly allocating the technologies of a company you know into these four cells as an exercise.

When applying the technology-review approach to a particular project within the company, the following typical procedure might be considered. It is based around a design created for use within a larger organization.

1. Obtain an objective view by selecting an audit team comprising of closely involved personnel and people who are not directly affected by it.
2. Have the project manager present technology and R&D strategies to the team.
3. Have the project manager identify key strategic areas and technologies, stressing issues that may affect these strategical areas and technologies and the skills needed to manage them.
4. Interview staff, managers, customers, and suppliers to obtain the internal and external perception of the company's technological stance. Summarize the findings in a clear statistical manner (e.g., a bar chart).

5. Where applicable, quantitatively measure the company's technological developments (e.g., patents, new products, scientific papers). Obtain a summary of the company's technological stance by identifying successes and failures, listing the factors responsible for each.

6. Identify organizational mechanisms that are used or are expected to be used to transfer technology (e.g., the R&D team, product development staff). The processes may be summarized by using a flow chart.

7. Review all of the team's information by using rating sheets for each project and program. Rate the requirements of the new technology against past developments, technological requirements, and fit within the organizational mechanisms to obtain a prioritized list of projects.

8. Analyze the results and draw conclusions for the technology strategy and the probability of success of each project.

9. Present findings, make recommendations, and follow up.

Thus, a project review (what we want out of the "what we could have" options), a competitors' review (what they've got), and a state-of-the-art technology review (the "what we could have" options) are obtained.

Here is a final overall checklist that a review should ask of any organization:

- What is the current situation?
  - What are the key technologies and know-how on which the business depends?
  - What is the company's status in these technologies? Does it lead or follow its competitors? What technology may be developing outside which may adversely affect the current situation in the market?
  - How did the company acquire these technologies? Were they made in-house or brought in?
  - Have we looked at everything to do with our current technology? Are there no new things we could do with it?
  - How do the company and its existing products compare with its customers' expectations?
  - How much longer is the current technology going to last?
  - What processes and policies are in place to identify product life?
  - What relative technological strengths and weaknesses are there in comparison to the competitors? Are there some products or technologies held onto merely for historical reasons?
  - What presently drives technological management? Quick fixes? Operational profit? Strategic considerations?
- What does the company intend to do?
  - What is the proposal for the new technology?
  - Can the company sell the existing technology and gain from being "ahead of the game"?

- Has the company optimized its exploitation of the technologies beyond integration into products? Has it maximized the technologies through strategic alliances, licensing, joint ventures, or cooperative R&D?
- Have strategic alliances been developed to obtain basic or distinctive technologies?
- What can we make the situation become?
  - How will continuing with the new technology affect the company's status in the market? Will it enhance differentiation? Technological lead? Product or service uniqueness? First-mover advantages?
  - Has this sort of thing ever been done before? If so, what is the track record, or what can be learned from the previous experience(s)?
  - How effective is internal transfer of technology? What communication networks are in place? Are they formal or informal?
  - Have the barriers to effective transfer of information been identified and removed?
  - Are the technical personnel available to fully exploit the technological opportunities?
  - Is there a process in place to integrate the technology and strategic business planning? If so, how effective is it?
- Moving on to the next stage...
  - Is the full support of all of the management of the company in place? This is a key milestone in achieving the goal of the new technology.
  - Does the company fully believe in the technology and its success?
  - Have the technology audits been effective in highlighting areas not previously covered?

3.5 ESTABLISHING MARKET POTENTIAL

Information required for the business plan and feasibility study includes the establishment of the market potential. In Chapter 2, it was shown that market research was necessary to make informed decisions on the options for the company. This process should be updated at this stage of the technology transfer process, particularly because new developments can cause the market to change dramatically, and often quite quickly. There are two main techniques involved: scenario planning and market research.

With scenario planning, a description of the possible future (a scenario) is produced, based on certain assumptions. It allows a chance for decisionmakers to assess their reactions to possible future changes. Scenario planning will be covered in detail in Chapter 11 on minimizing the risk of failure for the technology-transfer process.

Market research involves collecting customers, recording data on competitors, classifying other influences, and analyzing the results. The data on customers will arise from discovering both the overall market size and the needs of the customers. Techniques involve:

- Desk research (e.g., the Mintel database on growth in market sectors);
- Field research, such as face-to-face interviews, telephone interviews, mail surveys, panel discussions of interested people, and questionnaires based on defined key objectives and statistically valid samples.

In addition, market research determines what segments of the market exist and seeks to define the product in customer terms (features, benefits, proofs, who will buy the product first?).

Data on competitors is obtained in an altogether different manner. A current vogue expression is *benchmarking*, whereby company characteristics—its strengths and weaknesses—are compared with those of a competitor and the differences are then studied.

The primary use of benchmarking in this chapter is to identify areas of strength to further develop and areas of weakness to remedy or withdraw from. Benchmarking, however, is also an extremely important technique for determining planning objectives and improving quality by incorporating best industry practice into the company.

3.6 BENCHMARKING

Benchmarking was pioneered by Xerox. The company best known for its photocopying machines was losing both market share and profitability during the late 1970s, so its managers authorized a study of the quality, features, and unit costs of the competition's machines in comparison with Xerox machines. They were shocked by the initial results. Xerox's unit manufacturing cost equaled its Japanese rivals' selling price in the U.S.; Xerox had nine times the number of production suppliers than did the best companies; assembly-line rejects were ten times higher; product lead times were two times longer; and defects per 100 machines were seven times higher. Although Xerox executives could not believe the results, new benchmarking studies confirmed the data. The benchmarking process led to a meeting in 1983 involving the top 25 managers in the company to plan a quality strategy. The benchmarking results helped them to understand the amount of change that was required and to set realistic targets to guide their planning efforts. As example of adopting best industrial practice through benchmarking, Xerox used an adaptation of the warehousing and distribution practices of the L.L. Bean retail group (benchmarking should not be limited solely to direct competitors if research resources allow) [3].

Other benefits of benchmarking include:

- The process is motivating. It provides targets that have been achieved by others.
- Resistance to change may be lessened if ideas for improvement come from other industries.
- Technical breakthroughs from other industries that may be useful can be identified early.
- The process broadens people's experience base and increases knowledge.

A company that makes the same product is not necessarily a competitor. Companies with similar products may be selling to entirely different markets. Conversely, competitors are not always immediately obvious. One product can in fact be a substitute for another. Finding financial details (e.g., size or profitability) of competitors may be difficult. In Britain, for example, reports submitted by law to the Companies House Library may be incomplete, inadequate, or of no value.

Directories available from local libraries, however, can yield useful indicators. For example, if staff numbers are known, the payroll can be estimated from industry norms (allowing for knowledge of competitor attitude). From this, turnover can be estimated using standard operating ratios.

Important competitor characteristics required by a business plan are:

- Description;
- Size;
- Profitability;
- Operating methods.

Analyzing these characteristics can yield:

1. Knowledge of competitor weaknesses and their likely reactions to your innovations;
2. A definition of your product's points of difference (based on key factors of success in your sector).

Further information on competitors can be found from:

- Buying/analyzing their products;
- Studying their advertisements;
- Visiting trade exhibitions.

Product, service, and process improvements can only take place in relation to established standards, with the improvements then being incorporated into the new standards. Benchmarking measures an organization's operations, products, and services against those of its competitors in a ruthless fashion. It is a means by which targets, priorities, and operations that will lead to competitive advantage can be established.

Put another way, benchmarking is the continuous process of measuring products, services, and processes against those of industry leaders or the toughest competitors. This results in a search for best practices—those that will lead to superior performance—through measuring, continuously implementing change, and emulating the best.

There may be many reasons for carrying out benchmarking. Some of them are set against various objectives in Table 3.1.

Table 3.1
Reasons for Benchmarking

Objectives	Without Benchmarking	With Benchmarking
Becoming competitive	• Internally focused • Evolutionary change	• Understanding of competitors • Ideas from proven practices
Industry-best practices	• Few solutions • Frantic catch-up activity	• Many options • Superior performance
Defining customer requirements	• Based on perception • Based on history or gut feeling	• Market reality • Objective evaluation
Establishing effective goals and objectives	• Lacking external focus • Reactive	• Credible; unarguable • Proactive
Developing true measures of productivity	• Pursuing pet projects • Route of least resistance • Strengths and weaknesses not understood	• Solving real problems • Understanding outputs • Based on industry-best practices

There are four basic types of benchmarking:

- *Internal:* a comparison of internal operations;
- *Competitive:* specific competitor-to-competitor comparisons for a product or function of interest;
- *Functional:* comparisons to similar functions within the same broad industry or to industry leaders;

- *Generic:* comparisons of business processes or functions that are very similar, regardless of the industry.

The evolution of benchmarking in an organization is likely to progress through four focuses. Initially, attention will be concentrated on competitive products or services, including, for example, design, development, and operational features. This should develop into a focus on industry-best practices and may include, for example, aspects of distribution or service. The real breakthrough is when the organization focuses on all aspects of the total business performance, across all functions and aspects, and addresses current and projected performance gaps. This should lead to the final focus on true continuous improvement.

At its simplest, competitive benchmarking, the most common form, requires every department to examine itself against its counterpart in the best competing companies. This includes a scrutiny of all aspects of their activities. Benchmarking that may be important for customer satisfaction, for example, might include:

- Product or service consistency;
- Correct and on-time delivery;
- Speed of response or new product development;
- Correct billing.

For impact, the benchmarks may be:

- Waste, rejects, or errors;
- Inventory levels/work in progress;
- Costs of operation;
- Staff turnover.

The task is to work out what has to be done to better the competition's performance in each of the chosen areas.

The process has 15 stages, all of which are focused on trying to measure comparisons of competitiveness:

- Plan:
  - Select departments or process groups for benchmarking.
  - Identify the best competitor, perhaps using customer feedback or industry observers.
  - Identify benchmarks.
  - Bring together the appropriate team.
  - Decide information and data-collection methodology (do not forget desk research!).
  - Prepare for any visits and interact with target organizations.
  - Use data-collection methodology.

- Analyze:
  - Compare the organization and its competitors, using the benchmark data.
  - Catalog the information and create a *competency center*.
  - Understand the *enabling processes* as well as the performance measures.
- Develop:
  - Set new performance level objectives/standards.
  - Develop action plans to achieve goals and integrate them into the organization.
- Improve:
  - Implement specific actions and integrate them into the business processes.
- Review:
  - Monitor the results and improvements.
  - Review the benchmarks and the ongoing relationship with the target organization.

An alternative six-step process used by Alcoa in the U.S. is:

1. Decide what to benchmark. The following questions are often asked to guide the activities of the benchmarking team: Is the topic important to the customers? Is the topic consistent with Alcoa's mission, values, and milestones? Does the topic reflect an important business need? Is the topic significant in terms of costs or key nonfinancial indicators? Is the topic an area where additional information could influence plans and actions? The answers lead to the development of a purpose statement, which describes the topic to be benchmarked and guides subsequent activities.
2. Plan the benchmarking project. A team leader is chosen, who is responsible for seeing that the project is successfully completed. The leader should have the authority to make changes in processes, products, and services based on the benchmarking information. A team is then created, based on the range of skills needed for the project. The team's first task is to refine the purpose statement by considering questions such as: Who are the customers of the study? What is the scope of the study? What characteristics will be measured? What information is readily available?
3. Understand current performance. The team examines the factors that influence performance to learn which characteristics are most important and which are least important. Team members learn which data relate to the important characteristics and how to collect and measure those data. The collected performance data create the baseline and structure for benchmarking comparisons.
4. Study others. The team identifies benchmarking candidates, narrows the list down to a few candidates, prepares general and specific questions, decides the best way to get the questions answered, and performs the study.

5. Learn from the data. The team analyzes the data collected, quantifies performance gaps, and identifies which pieces of information might be particularly useful for improving performance.

6. Use the findings. The team works with the project sponsor to determine how the benchmarking findings can be best utilized and which other organizations in the company can benefit from its work.

Benchmarking is a multidimensional, multifunctional approach toward determining planning objectives and improving quality. To be effective, it must be applied to all facets of a business. For example, Motorola encourages everyone in the organization to ask [3] "Who is the best person in my own field and how might I use some of their techniques and characteristics to improve my own performance in order to be the best (executive, machine operator, chef, purchasing agent, etc.) in my 'class'?"

3.7 CONCLUSION

This chapter has demonstrated the vital importance of a systematic approach to information and communication. A good intelligence strategy can help you to learn from the best practice of competitors, avoid unpleasant surprises, find new technologies that are available (and cost effective), and ensure that research and internal communications work well.

References

[1]   Herring, J. P., "Scientific and Technical Intelligence," *Journal of Business Planning*, Vol. 14, No. 3, 1994, pp. 10–12.

[2]   Kuwahara et al., "Planning Research and Development at Hitachi," *Long Range Planning*, June 1989.

[3]   Evans, J. R., and W. M. Lindsay, *The Management and Control of Quality*, 2nd Ed., St. Paul, MN: West Publishing, 1993.

Chapter 4

Searching for the Technology

4.1 INTRODUCTION

Very few small to medium-sized companies will have the necessary expertise or research and development capabilities to provide new technology. Companies that wish to develop something new, therefore, must be able to find other organizations that can help them. Information is a key area in any search for new technology. Companies need to know where they can go to find information on products, research activities, finance, and legal aspects such as patents. This chapter looks at how research is transferred from academic institutions to commercial organizations. We look at initiatives around the world that attempt to help in this field. In particular, we examine one region in Germany that has set up a very interesting model for technology transfer. Finally, we take a brief look at the kind of information systems that are used in technology searches.

To set the scene for this chapter and to see how this cooperation between organizations works in practice, let's look at two typical examples. In the first case a research group needs a manufacturer, and in the second a manufacturer requires help from a university.

4.2 LINKS BETWEEN ORGANIZATIONS

4.2.1 Example A: Regional Medical Physics Department, Northern Regional Health, Newcastle-upon-Tyne, U.K.

Scientists and technologists at the Regional Medical Physics Department (RMPD) of the Northern Regional Health Authority research and develop medical instrumentation for the benefit of patient care. This is usually done in response to requirements from consultants in the region's hospitals. All of the instruments are highly precision engineered and fully developed for use in patient care.

The RMPD was under increasing pressure to earn revenue because the health authority was reorganizing. The department recognized the potential to generate funds through royalties by making some of these instruments commercially available through licensing agreements. One immediate problem, though, was that instruments developed as small batch prototypes were often not value engineered (i.e., they would need to be redesigned for cost-effective mass production). They would, therefore, have to spend time researching the use of cheaper parts and adhesives and development methods. Thus, they asked the local regional technology center (RTC) to identify suitable companies to manufacture and market selected instruments that were considered to have the greatest market potential.

One of the chosen instruments was the Neurotonic Therapy System, an instrument designed to give psychiatrists far greater control of the eletroconvulsive therapy used in the treatment of depression. The RTC identified Sycopel Scientific as a potential partner to manufacture the Neurotonic Therapy System. Sycopel was an established manufacturer of scientific instrumentation. The RTC helped in the discussions and negotiated a licensing agreement between the two companies. The RMPD was delighted to find a company with an excellent reputation. The company would manufacture the system. In addition it would offer a calibration service to customers and provide a replacement machine while the calibration work was being carried out. Sycopel was also delighted with the partnership because the machine had been expertly designed and met all international standards, and the company's market research showed that it had the potential to help hundreds of patients. They expected a large worldwide interest in the Neurotonic Therapy System.

This is a very good example of two organizations needing to come together for the profit of both. The research institute needed the commercial arm; and the company needed a new product. This example also illustrated the role of a *broker*, in this case the RTC, acting as a source for the medical institute and a provider for the manufacturer. The role of brokers like the RTC will be discussed in more detail later in this chapter.

There are of course numerous routes by which an innovation can come into contact with a producer. A different route is shown in our next example.

4.2.2 Example B: Almhults EL Mek, Sweden

Almhults EL Mek was established in Sweden in 1955. The company was originally a plant for the service and reparation of electrical engines and industrial machinery. In the early 1970s the company started repairing lifting magnets. The makes that were most common were Demag (of Germany), Steinert (of Germany), and Boxmag and Birmingham (of Britain). These were all heavy and expensive.

The company decided to develop a magnet of its own. The aim was to create a highly effective magnet that was lighter than those made by the competitors. These qualities would be suitable for mobile sites in particular. Other characteristics required of the magnets were high security and simplicity in use. The company, though, did not have the research expertise for this development, so it turned to universities for help. The project was taken up by the University of Vaxjo, which developed the lifting magnet. The university also helped with market research and interviews to specify which qualities the potential customers expected.

This example illustrates that universities can and do help companies. However, with the technology-transfer process in mind, it is worth noting the subsequent changes in the company. Towards the end of the development phase of the magnets, the company was doing authentic testing in three different customer companies. These companies were analyzing and measuring the usage. The results were used to make the final improvements to the product. Customer testing meant the product was developed to the needs of the customer. The development time was also used to market potential customers using the experience gained from the test customers. The company has gone on to develop three more magnet products in the separation fields, two of which separate nonmagnetic materials from each other by magnetic sound. The company has moved a long way from being a mere repairer of machinery.

These are two examples of organizations forming a partnership to enhance themselves and to transfer technology. Where does one look though for partners or for the sourcing of technology? How can research institutions transfer their research to commercial organizations?

4.3 USING HIGHER EDUCATION RESEARCH AND DEVELOPMENT

We have seen in Example B in Section 4.2.2 how a university can help a company with a new opportunity, particularly in the area of research and development. Companies are successfully using partnerships with higher education (HE) establishments to:

- Access new technology;
- Keep abreast of new developments;
- Access consultancy skills;
- Develop new technology jointly.

Also, HE establishments can generally provide research facilities at a much cheaper rate than a commercial company can.

Consider again the example of Seabait provided in Chapter 1. This is an extremely good case of how the technology-transfer process helped in the setting up and growth of a small company. The company was formed as a result of research carried out at the university. The links continued after the formation of the company, which in fact proved a lifesaver for Seabait. When diseases struck the rag worms in the early days of the company, the continued link with the university research facilities and expertise proved invaluable. The company solved the problems in the rag worms to avoid commercial disaster.

Thus, HE establishments can offer good value regarding technical expertise, and they can provide the availability of equipment beyond the means of any small company. One disadvantage of HE establishments is that they do not always fully appreciate the commercial aspects and concerns of the industrialist. This can be particularly true concerning timescales because academic research does not function with the same degree of urgency as does industry. There are different priorities and practices in HE in comparison with industry. This is particularly so with regard to the policy toward intellectual property in HE. Far more emphasis is placed on the publication of research than producing inventions for patents. It is widely accepted that care should be taken when making links between HE and industry. The best possible match with the technology must be found. For instance, with new materials such as electronic ceramics, experts may exist in HE in electronics or ceramics but not in electronic ceramics. It is then necessary to decide whether electronics experts could pick up the nuances of ceramics quicker and more effectively than ceramics experts could come to terms with electronics.

Much research is carried out in universities, but by no means is all of it for practical or commercial use. In fact, a useful statistic, if it was available, would be the percentage of research that classified as having some degree of commercial viability. A major problem for industry is knowing what research is going on in the university sector. This is due to lack of information, knowledge, and—in many companies—time to spend investigating. The companies also have to know what they want. The gap between the type of research carried out in universities and the companies' prediction of their future needs is the stumbling block in the links between the two. The potential is there—and this potential is what governments see as a major catalyst for forming new companies to apply the research from the public sector. They hope these links will provide many jobs for the people and profits for the companies and will create new wealth for the country. However, the links between HE and industry are very tenuous and vary from region to region and country to country. Such links must be beneficial to both universities and businesses—but perhaps this is where the main problem occurs. In many countries there are no benefits, status increases, or rewards for university staff forming links with industrial organizations. Like all partnerships, the relationships need to be beneficial to both parties if it is to work and function satisfactorily.

The following example shows how an idea for a product that could not be made at one time because the technology was not available eventually made it to the marketplace. This was achieved because pure research at universities later provided the technology that could be used to rejuvenate the invention.

4.3.1 Example C: The Artificial Nose, University of Manchester Institute of Science and Technology, U.K.

Many commercial processes, such as the production of tea and beer, rely on the experience of human personnel for quality inspection. These people develop a *trained* smell and taste for a good product. And, because the sense of smell is an essential safety device in some environments, and other environments are not desirable places for humans to work, both researchers and industrialists have identified the need for a nonhuman smelling device. This has now been achieved after joint research between personnel at the University of Manchester Institute of Science and Technology (UMIST) and the University of Pisa. The "nose" is based around an array of sensors formed with semiconducting polymers. The nose detects odor or gas to which the sensors respond, and this information is compared with that stored in the memory of a computer. There are advantages in using an electronic device over human ability in that the device can be used at ambient temperatures, it is quick, and it can recognize complex mixtures. A more important aspect is that it is not susceptible to poisons. The market for such a product includes the food industry, food storage companies, air conditioning, and water-quality monitoring.

The original idea for this device was conceived as long ago as the 1960s, but the necessary technology was not available at that time to make much progress. It was only in the 1980s, when research papers were being produced on conducting polymers, that the idea was regenerated. The first invention came in 1985, but it was not until 1990 that it was at a sufficient production state to begin commercial appraisal. In 1994 the product was ready for the market. It is noteworthy that the product was entered for the BBC/Prince of Wales Award for Innovation in 1995 and actually achieved a place in the final six entries.

4.4 LINKING ORGANIZATIONS TO EDUCATION ESTABLISHMENTS

Technology transfer from university to industry is clearly not one single operation. There are many combinations of methods being used to provide different outcomes. A factor often forgotten is that a major way of transferring technology is via people. The most general ways of forming links are as follows:

1. *Graduate employment.* Many large companies employ graduates as soon as they leave college. They believe that certain jobs require the expertise of a person qualified to degree level. There are many debates about the quality of graduates and whether universities train them sufficiently for industry, but that is not the concern of the theme of this chapter. SMEs tend not to employ graduates on the basis that they will not be able to pay them sufficient wages and that a graduate is more likely to move on to another company. There is also a feeling that the job level and interest may not

be sufficient for a graduate. SMEs could benefit, though, from the transfer of skills of a person who has been through a university system. Some countries have realized this and provide grants to companies to subsidize the wages of a graduate in the first two years. This seems a very good way of increasing the transfer of technical skills and knowledge to SMEs.

On a similar front, some countries have what is often known as a *teaching scheme*. This is a partnership between a university and a company to allow a graduate or several graduates to work on a specific project of technology in the company. This comes at little cost to the company, and the normal period for the project for any one graduate is two years. The project must benefit the company, but there must also be supervision and training of the graduate. This means the graduate gains considerable experience in an industrial setting. This again aids the technology-transfer process of a country, and all parties—the university, the company, and the graduate—gain from the scheme. In France, there is a scheme known as the convention on higher technical research (CORTECHS). This scheme brings together three partners:

- A higher grade technician with either a degree from a technical university or a high school diploma plus two years of training;
- A company that is prepared to employ the technician, either on a short-term contract or permanently, to work on an innovative project;
- A center of excellence to coordinate the project.

A lump sum is paid to the company by the center of excellence, on behalf of the Ministry for Research. This payment covers half of the salary costs of the technician, including insurance contributions. The technicians must be given 10 days of training in technological development and innovation management. So far, of those technicians employed via this scheme on short-term contracts, 94% have remained with their company on a permanent basis. It was also found that it was the first time that most of the companies had recruited a high school graduate with two years of training.

2. *Sabbaticals.* A sabbatical enables university lecturers to work in a company and thus gain insight into relevant industrial research and activities. At the same time, this allows the company to gain from the expertise of the lecturer—knowledge that may not be available inhouse. A sabbatical can serve as a stimulus for ideas both for the company and the university and could lead to joint research projects.

3. *Industry/university research units.* Organized research units have been proven to be highly successful in transferring research to industry. An example of such a unit is one that is an engineering design center for the transfer of individually made products to the marine industry. The units

are generally housed in the university, but they facilitate the formation of consortia with industry to work together. The benefits for the companies include early access to applicable research results, close interaction with university researchers and graduate students, joint research projects, and the provision of a network of organizations with similar interests.

Many countries have schemes to encourage collaborative postgraduate studentships (e.g., CASE awards in the U.K.). The research, generally leading toward a doctorate, is carried out in line with company needs and with company cofunding.

4. *University/industry liaison units.* Most universities realize the need to provide industry with the results of appropriate research projects. They have learned from the past, when a lot of new technologies developed in a university were simply given away to industry. As a result, each university has a campus-based commercial technology-transfer unit that organizes the protection and development of valuable new technology and makes it available to industry. This provides income for the university and staff and aids industry in finding new products. The professionalism of such units is debatable, however, and varies to a great extent. A major factor in their success is the status the unit and/or staff have within the university. Sometimes the unit is not given much credence in the organization of the university, and generally in this case the success rate is very low. The experience of the person in charge of the unit is also a major factor in its chances for success. The job requires a variety of capacities that includes good communication skills, technical and marketing expertise, industrial experience, and the ability to present a very professional image of academics. These qualities are not easy to find in the academic world.

5. *Forums for the exchange of information.* A proactive university interested in the success of industrial partnerships can prolong and sustain these partnerships by organizing networks of companies to exchange knowledge. The means of exchange could be via newsletter, workshops, seminars, and short courses. The formation of industry clubs is a very useful way to promote good long-term relationships with industry. The clubs hold regular meetings and organize special presentations

Vast improvements are being made in the way academic institutions work with the industrial world. Initiatives such as those in the European Union (EU) Framework program (see the appendixes) have helped to further the links. Further improvements are still necessary if we are to harness the expertise in the academic sector.

4.5 OVERVIEW OF RESEARCH AND DEVELOPMENT

4.5.1 Global Perspective

Seven countries account for the vast majority of R&D carried out in the main market economies. In order of expenditures, these countries are the U.S., Japan, Germany, France, the U.K., Italy, and Sweden. Between 1980 and 1989, although the U.S. was highest in research expenditure, the growth of spending for the seven countries was: Italy, 110%; Japan, 97%; France, 51%; U.S., 51%; Sweden, 44%; Germany, 41%; U.K., 8%. About 50% of all research expenditure in the seven countries comes from industry, with a significantly larger percentage in Japan and quite high percentages in Germany and Sweden. Universities are responsible for about 10%–30% of all the research conducted. This percentage takes into account both private and public funding of research in the HE establishments. The growth in the importance of university research has been more significant in the U.S., the U.K., and Sweden, with a decline in importance in the other countries.

The U.S. leads in the number of patents issued annually, which illustrates the importance of the U.S. in the world economy. More than 40% of these are filed by foreign inventors. Foreign inventors also account for a significant proportion of patents filed in other important markets—more than 80% in the U.K. and Canada and more than 60% in Germany and France. In Japan generally less than 20% of patents go to foreign inventors. The final piece of data on changes is the growth in high-technology manufacturing. Between 1980 and 1990, the U.S. growth was 10.4%; Japan, 19%; and the EU, 4.8%. The gains in the high-tech market for Japan were at the expense of the U.S. and the EU.

4.5.2 Japan

In Japan there exists a very strong desire and potential, particularly in industry, for active R&D. The main requirement of the university sector has been to produce highly educated individuals suitable for use in industry rather than research output. Industry's share in national R&D expenditure is about 80% and of the government's input of 20%, only 9% goes to universities (the rest goes to research institutes). This means that the level of R&D in Japanese industry is extremely high, while that of the public sector is the lowest in comparison with other countries. In Japan the Ministry of International Trade and Industry (MITI) is responsible for industrial technology policy. It is quite amazing that MITI has been able to stimulate industrial R&D very effectively with a small budget. The policies of MITI are not just formulated by government but also through open discussions of councils, such as the Industrial Structure Council and the Industrial Technology Council. The councils are composed of representatives from a variety of sources in addition to industry, such as finance, academia, journalism, small businesses, consumers, general workers, and local public authorities. Only a limited amount of government-funded R&D takes place. The aim is the solution of national problems such as the energy problem. Japan is able to reconcile private and public interests and capitalize on the dynamism of the private sector because MITI and industry meet each other halfway. MITI sees the promotion of business interests as benefiting the nation, and private firms are willing to make allowances for collective interests. Therefore, MITI and industry are able to avoid the confrontations that seem to affect government-industry ties in Europe and the U.S. MITI's ability to administer industrial policy is greatly facilitated by the organizational structure of Japan's private sector. This is specifically the industrial associations and business foundations, company specialization, *keiretsu* structures of clusters of companies, close banking-industry ties, and extensive patterns of intercorporate stockholding.

The contrast of the R&D policy and practice in Japan is so different from those of Western countries that it is quite fascinating and certainly worth further study. If you are interested in this aspect, we recommend that you look through the books about Japanese R&D and industry noted in the bibliography at the end of this chapter. It is also worth comparing Japanese practice with the problems of industry in the U.S. by reading the book *Made in America*, also cited in the bibliography.

4.5.3 European Union Research Programs

It is clear from the research programs of the EU, known as the Fourth Framework, and publications from governments of member states that research is not only a key component of economic growth, it is essential. However, for economic prosperity, that research must be coupled with industry. The EU has allocated just over 12 billion ECUs to the Fourth Framework program, to run from 1994 to 1998 (see Appendix A). The Framework programs cover the entire range of Union research, technological development, and demonstration. The mission of the program is that research can and must play a vital role in promoting growth and improving the quality of life in the Union, by increasing knowledge and skills and by virtue of its economic spinoffs. Compared with its competitors, Union research has been underfunded. The Framework funding is intended to counteract that decline through four activities. The first and second activities are related directly to research within the EU and with third countries and international organizations. The third activity, the dissemination and utilization of (research) results, and the fourth activity, the stimulation of training, are directly related to technology transfer. Further details of the activities are given in Appendix B.

There are already a number of EU programs in place designed to foster the commercial use of research, and details of these are given in Appendix C.

We would like to stress the following reminder about our extended use of the EU research program in this book. There is a large emphasis in this chapter on EU programs, and the appendixes contain very full details of these. We are certainly not trying to claim or show that the EU has the best system. Neither are we intentionally showing bias to the EU because we reside in one of the 15 member states. In fact, many of the aims of the EU programs are to catch up with other groups or countries, such as the U.S. or Japan. We are more concerned with citing the programs as examples of the present thinking on innovation and technology. The EU program provides a very good example of what a large partnership can provide in the way of policy or funding for R&D.

4.6 SCIENCE PARKS

Science parks are used all over the world as a means of helping small to medium-sized businesses. Over 450 science parks are currently in operation all over the world, 170 of which have locations in Europe. As one of the support programs for innovation, the EU has a science park consultancy scheme under the SPRINT initiative (further details of EU programs are provided in Appendix C).

The science park is a venture that brings together several complementary groups in an effort to help them grow as technology-based enterprises. A number of interest groups normally form the basis of the partnerships in the parks. They include universities, development organizations associated with the universities, small technology-based companies, and R&D sections of large companies. Often also on site are financial backers of the park, corporate sponsors, local government agencies, and government advisory bodies.

A development from the science park idea in Europe is the European Business and Innovation Center Network (EBN). Set up by the EU in 1984, this supports the needs of small new companies in training and services. Parks that join the scheme are helped with the establishment of an information center that will provide some of the services directly, but which has access through an electronic network to information held by other centers. This allows the center to act as a matching point for joint venture partners, as an agent for the procurement of technologies, and in other ways that are helpful to the organizations on the linked parks throughout the EU.

4.6.1 Example D: France

Sophia Antipolis, in the French Alpes Maritimes, is a science park on a grand scale. It is often said to be a technological utopia. It was established in 1969 by engineers and scientists from the Ecole des Mines in Paris (University of Mining), together with the regional government and local innovators. The development phase took 15 years, but, maybe surprisingly for a country with little foundation in provincial science and technology activity, it has proved very successful. It now houses many companies, including multinationals, research organizations, and laboratories of the University of Nice, L'Ecole des Mines, and Ceram. It employs over 5,000 workers and covers over 7,000 acres.

4.6.2 Example E: Greece

The PATRAS Technology Park was established in 1989. The mission of the park is to:

- Stimulate the development of new businesses based on technologies developing primarily within the University of Patras (established in 1964) and its research institutes as well as other Greek universities;
- Build on the interest in technology transfer of large Greek firms;
- Attract high-technology firms both within Greece and abroad to establish research and development departments for high-quality technological products and services;
- Encourage the industry to absorb technology originating in Greek universities;
- Create synergies with the research and development sector.

Seven technology areas were defined as being commercially exploitable that matched the core technologies in the University of Patras. These were information technology, telecommunications, electronics, materials, applied science, manufacturing, environment, and energy. The Greek financial system does not offer seed capital and is somewhat rigid on venture capital projects. Thus, one aim of the park is to alleviate this problem and help companies with initial funding.

4.6.3 Example F: Japan

Small business startups are rare in Japan, but the government is encouraging science/technology joint ventures in *science cities*. The first of these, Tsukuba, is well established. It was established with the move of a university out of Tokyo to a country site. The site was surrounded by major government research institutes, and large companies were encouraged to set up R&D centers on the periphery. The full effectiveness of this project is impossible to assess yet. The site has attracted foreign companies, and the model has created great enthusiasm in Japan. Further cities are now being developed.

4.6.4 Example G: Sweden

The Novum Research Park was opened in 1990. The main activities of the companies situated in the park are in developing products for the healthcare market. These companies have close links with the Karolinska Institute and Medical School, which has a research institute on the park. Also situated on the park is a government organization that tests medical products and assists the health services in their procurement policies.

4.7 AGENCIES

A further mechanism used for the promotion of technology is *technology transfer agencies*. The extent of the services of any agency varies. The types of things offered are searches for information on products and patents, patent and legal advice, and consultancy. Most of the agencies offer to find clients for the sale or license of a technology in return for a set fee or for a percentage of the return. Some agencies are contracted on a retainer fee by state or regional organizations or by private companies to find relevant technologies. These agencies can be very useful, as some specialize either within a specific industry sector or within a certain country or region.

The EU has been promoting the development of technology transfer networks over the past few years with the objective of improving the processes by which technologies available in one part of the EU may be made known to, and used by, enterprises in other regions. The European Association for the Transfer of Technologies, Innovation, and Industrial Information (TII) is an association of technology suppliers and brokers. Its aim is to stimulate innovation and promote technology transfer in the EU. Suppliers include R&D organizations, technology consultants, universities, industrial research associations, and manufacturing and service firms. It publishes a directory of its members with names and addresses classified by country, activity, clientele, and specialization. It also offers a range of ancillary services to its members, including a newsletter, information service, and conference organization.

Another EU initiative is the development of value relay centers in each of the member states. This network of centers became operational on January 1, 1993. The service is designed to encourage participation in Union research and technological development programs by industry, particularly small to medium-sized organizations. The relay centers offer their services to every form of enterprise—small or large, public or private—and to information brokers, consultancies, and finance organizations. These services are:

- Comprehensive information on EU programs;
- Investigation of possibilities to participate in R&D programs and advice on procedures;
- Advice on the possibilities of using the research results of their own company or scientific institution in cooperation with other interested parties;
- Quick access to information on applicable R&D results;
- Contact of networks and databases to support the search for partners for research projects and to support presentations of R&D results to experts from industry and science;
- Practical advice during the running of research projects;
- Access to nonpublic information;

- Information on patents and other kinds of legal protection of R&D results;
- Information on funding possibilities.

In a similar fashion, more than 200 European information (Euro info) centers have been set up in all member states of the Union. The aims of the centers are to provide a decentralized means for small- and medium-sized firms to have access to EU information.

A Euro info center's typical information services include:

- Public announcements and calls for proposals;
- Legislation of the EU;
- Environment and consumer protection;
- Financing opportunities of the EU;
- Training activities and training programs;
- Eastern Europe policy of the EU.

A Euro info center's typical advisory services include:

- Project financing;
- Applications and project proposals to the directorates general;
- Market information.

A Euro info center's typical events services include:

- Seminars, workshops, and lectures;
- Information days;
- Trade exhibitions.

The difference between Euro info centers and the value relay centers is that Euro info centers mainly aim to *initiate* R&D activities while relay centers mainly aim to *utilize* R&D results.

In the U.K. there have been various establishments available to offer help with a technological idea. As with most countries, the U.K. government is continually striving to improve its policies and the advice services to industry. Thus, some would say "the goalposts are continually changing" and therefore it is difficult to give details of agencies that may no longer be around tomorrow. One group of agencies that is a good general example for us to look at is the regional technology advisory centers (RTCs).

In the 1980s it had become clear that Europe was losing a technological lead in many areas and manufacturing companies were becoming less competitive compared with U.S. and Far Eastern companies. In response to this problem, many European countries established RTC networks. The argument was that a regional approach was necessary to coordinate the resources required by innovation and to ensure that industry in all parts of the country was in touch with national programs and facilities. In the U.K. in 1987, the government initiated the establishment of such a network by providing funding for three years and then planning for the RTC to be self financing thereafter. The RTCs work closely with industry, educational institutions, and government departments acting as brokers in the technology-transfer process. One example of the work of an RTC was shown in Example A of Section 4.2.1, where the RTC, in Sunderland, U.K., was the source in joining together the Regional Medical Physics Department and the manufacturer.

In Europe other possible local advice resources are business innovation centers (BIC). The support for these centers comes from locally or regionally based funds. They vary in the types of assistance available according to regional requirements, budget restrictions, and staff availability. The idea is that the centers provide support for small firms or individuals to develop and commercialize innovative ideas. They can give advice similar to that of the RTCs, but they do not have the networking of the RTCs nor as full a range of expertise. There is the facility and the encouragement, though, for the centers to join the European BIC Network (EBN).

In the U.K., a recent initiative is the formation of Business Link funded by the Department of Trade and Industry. The Business Link program intends to provide a central contact point for *all* aspects of business advice and will direct inquiries to the most suitable organization. For example, inquiries of a technical nature can be referred to a local RTC. Business Link aims to use the most modern management and customer-inquiry databases available via a wide area network equipped with the latest computer technology and communications programs, such as desktop and video conferencing.

An often overlooked place for sources of information is a local patent information office. Patent office staff will search the patent databases find patents that fit a set of criteria as defined by the client. Each month these offices produce a set of abstracts. These can be used to look out for competitors' inventions or any patents that may fit a company's general development strategy. If an abstract does stimulate interest, a full patent description can be ordered. Incidentally, on applying for a patent full details of the invention must be disclosed, so anyone looking at the patent can acquire useful tips about the inventor's techniques. There was some interest a few years ago in the design of the hull of the first catamaran to compete in the America's Cup. All of the competitors and commentators were desperate to glimpse the catamaran and find out the design. Security guards were hired to keep inquisitors away and competitors hired divers to try to get the information on the keel. What they did not clearly know was that there was a patent on the design and thus was available, in full, for inspection by anyone!

One other new initiative in the U.K. is the Focus on Foresight program, which has come out of the government's Office for Science and Technology. Initially, the aim of the program is to:

- Put in place a process that will provide a strategic framework for assessing future global market needs that could stimulate wealth creation in U.K. industry;
- Suggest how this can be obtained;
- Indicate what technology and science is required to generate and sustain improved competitiveness.

This process will involve a new partnership between the science and engineering base, industry, and government to identify the drivers for change and to promote opportunities that emerge from science and technology. The output of the process will be information on global market and technology trends. Similar programs are in place in the Netherlands, Germany, the U.S., and Australia, and one has been in Japan for many years.

France has the National Agency for Enterprise Creation (ANCE). This agency has developed specific actions to support the creation and development of innovative firms in France. These actions include regular collaboration with the French Agency for Research Exploitation (ANVAR). ANVAR is an innovation organization with regional centers providing support at all levels for new technology-based companies. Each year it provides support for around 1,000 new or recent (less than three years old) innovative firms, in the form of subsidies and reimbursable grants. ANVAR also provides backing and monitoring for the Technology Dissemination Network, which now operates in at least 12 of the 24 French regions. ANCE provide feasibility studies and support for the formation of regional business networks. It participates in research studies on the creation and sustainability of innovative technology-based companies. Also, it takes part in the CHALLENGE+ program in France, which aims to train and advise innovative company founders and managers. A recent study by P. Mustar (Centre de Sociologie de l'Innovation, Paris) [1] highlighted the main characteristics of new company founders. These were quite interesting. They were:

- 72% are aged below 40 years, although the average age was 40.
- The founders possess a high level of technical and scientific qualifications.
- The founders also possess a high level of work experience, with an average of 10 years of managerial experience in most cases (research-based founders included).

Mustar also found data that consolidate findings in other countries, namely:

- The launch is difficult, the financial needs and skills required are demanding, the risks are high, and the commercialization phase is hard.
- All those who invent or who develop an innovation do not necessarily set up a company.

- Those that do start up a company develop and create jobs in a much greater proportion to other startup companies.
- Of those with which ANVAR was involved, 80% were still surviving and 32% had more than 20 employees after five years.

Most countries also have private companies whose main business role is as a technology broker. They work in a very similar fashion to the agencies already mentioned but raise their capital by other means such as the stock market. One such establishment in the U.K. is the British Technology Group (BTG), which is very prominent in the technology-transfer business in that country.

4.7.1 Example H: British Technology Group, U.K.

BTG has two current types of business. One is the development, under license, of technology from public-sector sources into industrial production. The public-sector sources are the universities and colleges (about 100), the research council establishments (about 100), and 15 government laboratories. The other business type is the provision of funds for innovative private-sector companies seeking to develop new products and process (usually as joint venture).

Once an invention is accepted for exploitation, BTG take responsibility for patenting, licensing, and policing the patent. BTG may decide to incur some costs in patenting, funding further development and negotiating license agreements to improve the prospects of successful international licensing. Any such license agreement would still be shared with the owner of the invention. (More information on intellectual property rights is provided in Chapter 9).

The professionalism and success of BTG in handling technology are widely recognized, both in the U.K. and internationally, and thus its business is expanding.

In the U.S., manufacturing technology centers (MTCs) were established in 1988. Their mission is to aid small- and medium-sized manufacturers to upgrade their technical performance and competitiveness. For six years, half of the funds for the MTCs were provided by the National Institute of Standards and Technology. The other half of the funds were to be found by the MTC. After six years, the MTCs were expected to be financially self sufficient. The aim is for the MTCs to concentrate on specific sectors appropriate to their region, such as textiles in South Carolina and advanced manufacturing in Cleveland. The MTCs are encouraged to attract company managers into the MTC rather than have agents visiting the company. MTCs are also encouraged to involve large manufacturing companies as technology-transfer partners, with original equipment manufacturers seen as the most powerful agents of change. This is a quite different perspective to that used in similar programs in Europe.

The U.S. also has set up regional technology transfer centers (RTTCs) funded by the Space Agency. The aim is to operate these as regional contact points for all industrial sectors, federal laboratories, and state government initiatives. The focal point for the program is at the National Technology Transfer Center in Wheeling, West Virginia. This center provides common training, database, and marketing facilities. Six RTTCs were set up in 1991 to act as industrial application centers. More recently President Clinton pledged $500 million to create a network of 170 MTCs. Interestingly, 170 is the same number of technology centers in Japan.

In Japan, regional support is provided through the 170 *kohsetsushi* centers. These target small companies with less than 300 employees offering a wide variety of innovation services. As was mentioned in Section 4.5, Japan's research is concentrated in industry rather than academic institutions. The large companies contribute to the innovation process by developing long-term relationships with their supplier networks. This is backed by resource sharing, personnel exchange, joint training programs, and advice on new systems and technology. Many studies show that by this process Japan maintains itself as the world leader in quality and sales in most manufacturing sectors.

4.8 THE BADEN-WÜRTTEMBERG MODEL

In Southern Germany, the state of Baden-Württemberg (BW) has the highest percentage of small- and medium-sized businesses in the country. About 95% of the companies in BW employ less than 500 workers, compared with a national average of around 90%. The region has also established a large proportion of world-class companies, such as Bosch, Porsche, Hewlett-Packard, and IBM. Thus, the area has the strongest regional economy in Germany. The strength of the economy is based on three main factors: dynamic networks between organizations; a rich institutional system of vocational training; and substantial public and private investment in R&D, innovation, and technology transfer.

The German government, at the end of the 1970s, placed in its policy a high priority on technology transfer. In 1982 a special technology-transfer model was created in BW. The state government of BW created the position of commissioner for technology transfer, a post directly responsible to the minister president, to oversee all aspects of this function. The commissioner obtains his know-how from the members of the Steinbeis Foundation for the Promotion of Economy. The foundation is a self-financing foundation of civil law, whose function is to assist local industry by giving advice on technological matters and by supporting firm development proposals. The commissioner also holds the post of chairperson of the foundation. The foundation has a main office in Stuttgart, and it functions as a coordinating and contact establishment for technology-oriented problems that are brought forward by industry.

To fulfill an efficient technology-transfer process, the foundation has formed a network of 100 transfer centers over the region. The network was formed around the existing infrastructure of the region. Most of the locations of the branch offices of the foundation are in close proximity to a technical college. This has enabled close cooperation with the staffs of these schools.

Between 1983 and 1989, the income for the foundation rose from DM 10 million to DM 65 million. About 90% of the budget comes from the 7% overhead paid by each of the transfer centers and about 5% comes from donations. The BW government provided a subsidy for opening new transfer centers, which in some years amounted to DM 2 million. A company using a transfer center is given the first five hours of consultancy free of charge, and each firm is allowed one free consultancy per year. This time is claimed back from the government at an agreed rate per consultancy. The foundation also undertakes special consultancies for the government (e.g., the government makes loans and grants available for investment in new technologies, and the foundation assesses the applications (2,000–3,000 per year) for a fixed fee per project). This is a very important point in that the BW government provides grants for companies so that they can pay for applied research at a university, technical college, or institute. In one year, the foundation will be involved in over 6,000 specialized consultancies, 4,000 R&D consultancies, 3,000 general consultancies, and 2,000 advice services. Over 70% of the business of the foundation is with SMEs employing less than 100 people.

The infrastructure for R&D in the BW region including:

- Higher/further educational institutions:
  - 9 universities, including two technical ones in Karlsruhe and Stuttgart;
  - 38 technical colleges;
  - 6 pedagogic high schools;
  - 7 art colleges.
- Public nonuniversity research institutions:
  - 14 Max Planck Institutes;
  - 3 large research organizations;
  - 14 Institutes of the Fraunhofer Society, a leading organization in applied research;
  - 6 foundations at the universities in the fields of microelectronics, data processing, laser technology, natural sciences and medicine, application-oriented processing of scientific data, solar energy, and hydrogen research;
  - 2 international research institutes;
  - 49 other nonuniversity research institutions.

All together, over 20,000 scientists and technologists work in the centers and institutes in the region. This gives BW the largest public R&D in the country. From the aspect of expenditure and employment, private funding substantially outweighs that of the public sector. Basic and long-term research is still in the province of the universities, public research centers, and Max Planck Institutes (which is entirely funded by the public: 90% by the federal government, 10% by the region one). The Fraunhofer institutes and research organizations are more involved in short-term applied R&D and testing. A key element in the technology transfer of research to industry is the work of the Fraunhofer institutes. The Fraunhofer Society was set up in 1949 as a framework for a number of existing independent research institutes. The aim of the institutes is to promote applied research mainly in the fields of natural and engineering science. They are funded 25% by public means and the rest by private means (although most of this comes from the public sector through competitive tendering).

There has been a move by universities in Germany to increase private funding for research. The University of Stuttgart now achieves 70% of its research budget from private, mainly industrial, sources. The Fraunhofer institutes undertake research on hardware development in all engineering fields, with an aim to match the knowledge of their staff with that of the industrial client. Companies contact the institutes to develop new products or processes using Fraunhofer technical equipment. The institutes bid against private companies or consultants for public contracts to develop a product or process. The research findings are then sold to the highest bidder for further technological development.

A very interesting and key concept is that the managers of the transfer centers are academics. These academics are have reached a senior level, such as a professor. They are carefully selected and they can keep or sell any patents they receive from an invention. The centers specialize in high-technology fields such as computer-aided manufacturing, automation engineering, optoelectronics, sensor technology, and quality assurance. All results from the work of a center are held on a database held at the Steinbeis Foundation, thus providing a common information resource. Another interesting statistic is that more than 1,200 professors and engineers work in the transfer centers. The foundation makes the investment for the equipment required, sometimes with government help. Importantly, the equipment of the transfer centers is used by the polytechnics, and polytechnic equipment is used in the transfer process.

In addition to the transfer centers, there are also technology centers. A center will house around 10 to 20 new companies. The location of the technology centers, like those for transfer centers, are in close proximity to universities or technical schools. There are 10 such centers in the region of BW. The benefit for the new companies is that they can get special technological consultations and assistance in management. Overall the main activity in all of the centers is research and development.

One other part of the support services in BW worth a mention are the *aninstitutes*, which translates to "beside institutes." These began working in 1986, and they are a further bridge between universities and industry. They are funded both by private and public means. The aim is for them to work quickly to meet the needs of technology transfer of the larger industrial companies. This is carried out by the signing of a legally binding contract, which includes delivery date and penalty clauses, to develop a technology for an organization. Aninstitutes exist at Karlsruhe, Stuutgart, Ulm, Tübingen, and Heidelberg.

So how does this help the companies? It is possible for innovators to use the laboratories and equipment of the universities and technical colleges for a small fee so that they can begin to work without too heavy an initial investment. They can form join projects with university staff. They can suggest topics for research in the university by students and staff. As a bonus of this system, more than 700 students work on projects, thus providing on-the-job training for them and solutions to problems for the industrialists.

In the product search and utilization of ideas, the staff of the centers can help the innovators in their efforts to diversify and look for suitable cooperative partners in industry to transform scientific ideas into concrete products and procedures. Those companies who wish to invest in another company get help in carrying out a technological evaluation of the company in question. Companies undertaking joint ventures with foreign firms get help in trying to find a partner who is already based in the BW region. In addition in settlement cases and joint ventures with foreign firms, the personnel of the BW Trade and Industry Office gives special consultation and assistance.

One of the most important business associations in Germany is the many chambers of industry and commerce (IHK). The chambers are affiliated with a central organization, the Deutscher Industrie und Handeltag. Unlike in other parts of the world, the chambers in Germany are public law corporations, and membership is compulsory for all industrial and commercial organizations. Each company in the country has to pay a certain proportion of its tax to its chamber. This means that the chambers can deliver a wide scope of high-quality services because they are so well resourced. The chambers play a vital role in education and training. Responsibility for management, standards, and examinations of apprentices lies with the chambers (e.g., the Stuttgart Chamber has four training centers specializing in metalwork and electronics). The chambers also organize special working groups in areas such as laser technology, microelectronics, data processing, and biotechnology. The knowledge from the groups is then disseminated to member organizations. The chambers act as innovation consultants via a system of innovation consultancy bureaus. There are eight such bureaus in the BW region. The chambers provide the following technology-transfer services:

- A primary gateway into a region's technology-transfer system for most SMEs, thus saving the SMEs the time-consuming task of researching the system themselves;
- The offer of basic technical consultancy, which, unless the chamber has to pay for some activity such as a database search, is offered free of charge;
- The use of their expertise and knowledge to find the most appropriate technology-transfer agency or partner for the member company.

The chambers in BW, therefore, are in some way in competition with the Steinbeis Foundation. The chambers have criticized the foundation for being too committed to the technical-school network (and thereby neglecting to some extent other technology-transfer agents, such as universities, Max Planck Institutes, the Fraunhofer institutes, and the private-consultancy sector). The main problem is that the transfer centers are, in turn, a major competitor to the private-consultancy companies. These private companies are members of the chambers and the transfer centers are seen to be subsidized by the state. It has not though proved to be a serious defect in the structure for innovation in the region. The chambers provide the resources to enable companies, particularly SMEs, to help themselves, and all of the companies are deeply embedded in the business networks of the chambers. This business network is highly developed in BW and is a particular strength of the region.

The region has been dominant in working with the European Commission in prompting interregional cooperation. It has formed links with Rhone-Alps and regions in Crete, Apulia, and Wales. The broad aims of cooperation are to assist companies in partner regions that are seeking locations or joint ventures in other parts of Europe.

The infrastructure for innovation in BW is certainly the envy of many other areas of Europe. The government has brought in new incentives for companies to collaborate at home and abroad. So, the region has continued to benefit because networking has become even more important. There is no doubt that many of the less favored regions in Europe can learn from the BW model. The public policy is successful in BW because it aims to help companies that are willing to help themselves. The high status given to product innovation, training, and new technology means that organizations are in a strong position to make the most of whatever support is available from the government. Finally, a new strength is emerging in the region—the experience and expertise of the larger companies is now being used to help and tutor the SMEs in a kind of mentoring network. The strong base from which the region is able to start and the continuing learning, and other improvements they are able to make to their system, mean that it is very difficult for other regions to ever match the BW paradigm.

For more details about the regional innovation system in BW, we refer you to Cooke's and Morgan's "The Regional Innovation System in Baden-Wurttemberg" and Herdzina's and Nolte's "Innovation-Infrastructure for Rural Economic Regions: Conception and Practice of Technology Transfer."

4.9 INFORMATION SYSTEMS AND SEARCHING FOR NEW TECHNOLOGIES

Most countries or unions of countries, such as the EU, will have information-technology systems to support private companies in their search for information. These systems will have databases that contain information on advanced-technology companies, technology-development projects, financial support to companies for technology development, and many other sources of data on innovation and technology. Many of the systems include telematic networks for reaching agreements on cooperation and technology transfer. There are many companies that market technological information databases on CD-ROMs. One of the first in the U.K. was British Expertise in Science and Technology (BEST), produced by the publisher Longman-Cartermill. Many of the very large companies and universities have computer-based networks that enable the access of information from worldwide sources. The databases and the technology are changing rapidly, but local technology centers such as the Business Link in the U.K., universities, or main libraries will be able to access the databases.

Let us look at, as an example, the systems supporting technology transfer in the EU.

The most notable databases in the EU are the Scientific and Technical Information Service (ARIST), CORDIS, and EPIPOS. ARIST is an information tool for locating existing technology in the market, and it is used before establishing contact with potential supplier companies or clients. It connects to all of the public and private databases and has information about patents and trademarks. This service offers companies a series of informational products classified into three groups:

- *Scientific and technical information.* This is information on the state of technology, which analyzes what stage the research of a particular technology has reached.
- *Technical/legal information.* This group offers complete references about subjects such as industrial ownership (patents, trademarks, use of models, national and foreign technical standards) as well as all the regulations and legislation affecting them.
- *Technical/economical information.* This offers market studies of supply, distribution, and means of supply of materials.

Information about the research activities supported by the EU is stored on the CORDIS databases. This project of provision of information was established in November 1990. One of the aims of providing the databases is to promote the exploitation of community-sponsored research results. At the moment there are eight CORDIS databases. The databases contain information on all EU-sponsored programs, project descriptions from all programs, and abstracts of reports and publications. It also gives the latest news on EU R&D activities plus results, partners, EU communications to the council and parliament, and contacts. CORDIS is a direct online system, but it is also available in CD-ROM format. This is updated on every three months. It also now has established a server on the World Wide Web allowing Internet users to access CORDIS through a graphical, hypermedia-based interface. Contacts for CORDIS can also be routed through the value relay centers mentioned in Section 4.7. A benefit of CORDIS is the ability to advertise your own research and development and any requirements for assistance you may have.

The European Patent Office produce a series of CD-ROM databases known as the European Patent Information and Documentation Systems (EPIDOS). This database gives patent information from over 50 countries. The system provides up-to-the-minute information on all major technologies. It enables users to learn about current developments in 95% of research worldwide and helps to point towards trends and areas of new significant activity. The EPIDOS system uses the information from other databases such as PATOLIS, the unique source of Japanese patent data, to provide a source of information that comprises over 50 million patents. Globally, new patent applications are filed at a rate of one every ten seconds. This means an annual total of 600,000 patents for Europe alone. EPIDOS databases are updated every day, which ensures that the information is always in line with current developments. EPIDOS offers other services, such as individual searches, patent document copying, translations of Japanese patent abstracts, and detailed statistical analyses of patent data. In its marketing, EPIDOS states it is "the gateway to world-wide information."

A major program in Europe has been and still is the EUREKA project, which will be mentioned again in Chapter 5. The EUREKA initiative, created in 1985, has 22 European countries and the EU as members. Its aim is to strengthen productivity and the competitive position of industry and national economies in the world market. EUREKA was established to promote technical collaboration in Europe and to encourage the joint development of "close to market" projects with clear commercial applications. A EUREKA database is available to provide information on the many projects and the great variety of partners involved in them.

Although the concentration has been on information systems in Europe, the databases mentioned are very typical of those used in other parts of the world, particularly in Japan and the U.S.

4.10 CONCLUSION

A tremendous amount of research goes on in universities, public research bodies, and industry. A patent is filed somewhere every ten seconds and this is only a fraction of the research being done throughout the world. This is the reason that governments are so eager to provide the right conditions for the exploitation of ideas. They see a mountain of research being carried out, and if the number of projects that can provide marketable goods is significantly increased, there is an increased chance that the policy of governments will be achievable. New jobs will be created, providing increased wealth for the nation. This sounds easy in theory but is very complex in practice.

A good deal of research is not applicable to commercial exploitation. The process of creating conditions that make the transfer from research institutes to industry has no magic formula. There are many parameters, and poor performance in any one area can cause a breakdown in the transfer behavior.

Two models stand out to study in any attempt to find the ideal solution. The Baden-Württemberg system, with the close relationship between academic institutions and industry, including the exchanging of staff, creates first-class conditions for the transfer of research. In addition, the Japanese thrust in forming very close relationships between suppliers in a manufacturing chain and between companies of different industrial sectors clearly has many good points. Either model would be costly to set up and may, in some countries, be difficult to achieve because of the transition required from present traditional methods. Perhaps the future lies in reproducing parts of such models on a smaller scale, or at least examining these systems and copying good practices from them.

References

[1] Engelbach, P., "The Creation of Innovative Firms In France," *Technology Transfer Practice in Europe Conference Papers*, Vol. II, European Association for the Transfer of Technology, Innovation, and Industrial information, Luxembourg, 1994.

Selected Bibliography

Andersen, C., *Getting European Community Help for Your Company*, London: Kogan Page, 1993.

Archer, C., *Organising Western Europe*, London: E.J. Arnold, 1990.

Dertouzos, M. L., *Made in America: Regaining the Competitive Edge*, Cambridge, MA: MIT Press, 1989.

Bower, D. J., *Company and Campus Partnership: Supporting Technology*, London: Routledge, 1992.

Charles, D. R., and J. Howells, *Technology Transfer in Europe : Public and Private Networks*, London: Belhaven Press, 1992.

Cooke, P., and K. Morgan, *Learning Through Networking: Regional innovation and the Lessons of Baden-Wurttemberg*, (RIR No 5), Department of City & Regional Planning, University of Wales, Cardiff, 1990.

Cooke, P., and K. Morgan, "The Regional Innovation System in Baden-Wurttemberg," *Int. J. Technology Management*, Vol. 9, No. 3/4, pp. 394–429.

Herdzina, K., and B. Nolte, "Innovation-Infrastructure for Rural Economic Regions: Conception and Practice of Technology Transfer," *Proc. T11, Technology Transfer in Practice*, Hanover, Germany, 1994.

Philips, G. O., *Innovation and Technology Transfer in Japan and Europe*, London: Routledge, 1989.

Watanhe et al., *The Inducing Power of Japanese Technological Innovation*, London: Pinter, 1991.

Chapter 5

Company-to-Company Transfers

5.1 INTRODUCTION

Chapter 4 emphasized the transfer of technology from institutions such as universities to companies. This involves transfers from the public to the private domain, as the research bodies' funds come from government sources. Research is not only carried out in public organizations, however. Private companies with research and development facilities produce innovative ideas that they may or may not patent or develop. The potential products may be outside their product range or considered to provide a return that is too small compared with development costs. These potential products may be suitable for another company to develop.

There are many ways in which organizations can become involved in helping other firms—hopefully, to the benefit of all parties. In the Seabait example in Chapter 1, you can see how helpful a large conglomerate such as Alcan can be to an innovative fledgling firm. Alcan was able to help a small company, not with any research but with its premises, by being a very generous landlord and doing whatever it could to help the small company through any difficulties. The payback for this large company is in public relations.

Company-to-company technology transfer will generally be rewarding for all concerned, but large company-to-SME *spinoff* is not easily undertaken. Generally, the most difficult part is to motivate the large company. It may be willing to offload an opportunity, but larger companies are often reluctant or unwilling to put effort into the development without a stake.

ICI is one large U.K. company, though, that is trying to help with redeployment of its employees into other companies. The scaling down of large companies like ICI might provide industry with an ideal opportunity for corporate venturing and technology transfer in general. In Europe, the oil company Elf Acquitaine promotes SME development as a feature of the corporate plan. This puts something back into the business community at a time of contraction within a region heavily dependent on the large company. The attitudes of large companies are changing as they realize they have responsibilities to communities in which they reside.

Many governments believe that the future prosperity of their country will depend on the degree and speed with which technological innovations are implemented. This implementation is more likely to be through smaller companies using the spinoff research and development from both public and private organizations. The success will therefore be very dependent on the way in which companies work together and the extent to which all members of the collaboration are able to gain advantages.

5.2 COLLABORATION BETWEEN COMPANIES

There are many definitions of collaboration. Collaboration between companies can also come under such titles as alliances, networks, and cooperatives. In industry, whatever the title, we are looking at agreements between organizations to work together. This could be with suppliers, customers, and even competitors. Many car manufacturers in recent years have formed alliances to work together yet retain their individual brand names. Honda of Japan collaborated for many years with the Rover Group in the U.K., and now BMW of Germany has taken over Rover. Collaborations can take a very wide diversity of forms in such areas as research, manufacturing, prototyping, product development, and marketing. They can, of course, be in all, some, or perhaps just in one of these areas.

We will concentrate more on the technical partnerships. One form of technology transfer in collaborations is where partners increase their expertise through sharing knowledge, skills, and personnel. In Japan, 75% of the Japanese car plants have customers who run suppliers' associations, and they in turn run associations for their suppliers. These associations take various forms but typically serve as places for information exchange and learning within the supply base of a car maker. Another very important form of transfer is where one company has the technology that another company needs for its new product. Yet another form is where a company takes on the technology from a collaborator and develops it into a product for the market.

So what are the aims of technical collaboration? The most important goal is to improve the innovation process. Public policy works to increase collaboration to improve the comparative technological performance of national organizations and enhance the national system of innovation. It is difficult to know the full extent of collaboration agreements around the world, mainly due to the lack of data. The Japanese high-technology industry has a large amount of R&D programs with collaboration and organizations. In the U.S., since the introduction of the 1984 Cooperative Research Act, which was brought in to protect companies from anticompetition laws, over 200 partnership agreements have been formed. The EU has actively promoted collaborations for many years through the Framework programs. One European (non-EU program) initiative named EUREKA (introduced in Chapter 4) encourages the formation of many industrial partnerships. These are in a wide variety of fields, such as medical and biotechnology, robotics and production automation, communications, environment, information technology, new materials, lasers, energy, and transportation. The aim of EUREKA is to strengthen productivity and the competitive position of industry. To obtain EUREKA status, projects must be a cooperation between participants (enterprises, research institutes) in more than one European country. They must also use advanced technology. The aim of the project must be to secure a significant technological advance in the product, process, and service concerned for which a viable international outlet exists.

This is an international scheme that is working very well. It must be stressed, though, that other collaborations need not be across country borders but could be just national or even regional.

5.3 WHY COLLABORATE?

Essentially, collaboration is necessary if a company is faced with a problem that it cannot solve by itself. This could be of a technical, commercial, or financial nature. It is impossible to put a figure on what percentage of collaborations fail, but it is not easy to make collaborations work. Shared activities clearly imply shared future profits and can also compromise technical know how. So why is it in the interest of companies to collaborate with other companies or research centers on technological innovation? The seven major reasons are:

- To share risks;
- To share costs;
- To gain technological knowledge;
- To help in product development;
- To develop industry standards together;
- To gain additional markets;
- To improve the speed to market.

New product development is generally costly. Even large companies may be reluctant to take on a new development because of the initial costs, and small concerns certainly have great difficulty financing new ventures. The development of a new product has often been disastrous for firms, even leading to bankruptcy. Thus, spreading risk and cost between companies is a safer approach for all concerned. Many companies are willing to take the gamble of producing a new product, but they need the help of a more established company to pass on technical expertise. This help can also lead to the product coming to the market a lot quicker. If we are to move forward with greater prosperity, as the politicians predict and desire, then there needs to be far more company to company partnerships. This will be a major factor for the growth of small- to medium-sized companies.

5.4 EXAMPLES

Let us look at some case studies of technical collaboration. The first example, from our own region of the U.K., shows the transfer of a technology product and the subsequent problems that arose. The next two are examples of technical collaboration in the EUREKA program.

The first example is also an example of how luck can play a part in transfers. The third-party agent happened to be in contact with two organizations on completely unrelated matters and realized that one had a product that the other needed. We have kept this example in the full version because it gives a further study of the full innovation and technology-transfer process and an example of company-to-company transfer. It is a case study to which we have experience, but it could be typical of companies anywhere in the world.

5.4.1 Example A: Hartnell Alarm Controls, U.K.

Hartnell Alarm Controls was established by Keith Hartnell in Wallsend during the early 1980s. The company gradually grew in size, supplying fire-protection systems for ships and marine diesel engine monitoring systems.

It became apparent that the amount of shipbuilding and repair work in the Northeast of England would decline rapidly and another market had to be found. In 1986, Hartnell was given a demonstration of a highly sensitive particle detector that had been originally developed in Australia. The detector, called VESDA, had been developed six years previously to monitor exhaust gases from engines. Hartnell saw the potential of VESDA, not in engine monitoring but as part of a high-performance fire-detection system.

Having secured a three-year distribution contract for the product, Hartnell set about developing a new market for VESDA in the U.K. The VESDA detector was around 1,000 times more sensitive than conventional smoke detectors. Although it was considerably more expensive, it was ideally suited for high-risk areas such as computer rooms. Early detection of fire was essential, as most computer-room fires are made safe if the power is shut off early. This avoids the usual method of extinguishing, using a costly and disruptive HALON gas dump. Early detection is difficult with conventional detectors, as the mechanical ventilation systems often found in computer rooms dilutes smoke before it reaches the detector.

As Keith Hartnell explained, "We developed a totally new market for an old product. The next problem was to find a new product for this market during the next three years." The search began.

Gordon Ollivere, director of the regional technology center, then located in Newcastle-upon-Tyne, introduced Hartnell to Techsearch, a database of technology-transfer opportunities. The search identified work on advanced particle monitoring systems at Hatfield Polytechnic. Unfortunately, further investigations proved that this technology was unsuitable for Hartnell. By chance, Gordon met someone from the Royal Signals Research Establishment (RSRE) in Malvern, part of the U.K. Ministry of Defence (MOD). RSRE had developed a laser-based particle measuring device with which to measure gas and exhaust flows in jet engines. Gordon suggested to Keith Hartnell that this technology might just be what he needed. Trials proved successful, a feasibility study was positive, and the search for a new product was complete. In 1986, Hartnell negotiated a technology agreement with the MOD to use its product. This was via the Defence Technology Enterprises, the technology-transfer agency of the MOD. The royalty agreed was around 10% of the selling price of the new detector.

The beginning of the next stage of the company's development was marked by a change of name to Hartnell Microelectronics. Hartnell had developed into a distributor and installer of high-sensitivity smoke-detection equipment, but it had no in house R&D or production facilities. Further investigations of the technology showed that although it was feasible, it needed a considerable amount of design and development to make it a reliable, accurate product that could be easily manufactured. The company needed to recruit more expert help. So Hartnell appointed a new technical director, an expert in optoelectronics, to oversee developments. Over the next two years, the development team would grow to about 10 staff. The company's aim was to launch the new HART detector at the industry's major sales exhibition (FIREX) in October 1989.

Besides the usual technical problems that had to be resolved, a serious problem also unfolded. The factory price of an individual HART detector turned out to be about 20% higher than that of the VESDA detector, despite continuous attempts to reduce it. The cause was primarily high component costs, but this was aggravated by the 10% royalty on the selling cost. Although the 10% on sales royalty does not sound particularly onerous, if the company is making a gross margin of 30%, the royalty is accounting for a third of what remains. A marketing strategy was needed to overcome this problem.

The HART detector had a number of technical advantages over the VESDA detector. Besides being more accurate and reliable, it required little maintenance, whereas the VESDA detector required frequent maintenance with expensive component replacements every two years. The control panel for the VESDA required additional modules for each extra detector connected. Each detector was wired directly to the control panel and, as fireproof cabling had to be used with fire-protection equipment, this was expensive. A new control panel was developed for the HART detector that could accommodate up to 64 detectors with no additional hardware. Digital communications allowed looped cabling, reducing installation costs. For installations above two detectors, the HART detector was competitive on price.

By not making certain parts of the HART system (i.e., detectors, control panel, and cabling design) similar in form to the original VESDA system, the customer could not make a direct comparison on individual component price. The marketing concentrated on overall costs of ownership, including installations and maintenance. This brought the customers' attention to functionality in which HART was more dominant than VESDA.

Coincidental with the launch was the move of the company from its original 2,000-sq. ft. premises to a 12,500-sq. ft. factory, which provided clean-room facilities and a large production hall.

By the time of the product launch it was becoming obvious that Hartnell needed an additional investment. This was to obtain necessary approvals, develop new markets, and establish volume production facilities to fully exploit the potential of the HART detector.

Williams Holdings, a large U.K. conglomerate, had been developing a market for the VESDA product in the U.S. through one of its American subsidiaries. It too was looking for another product for this vast market when the VESDA distribution agreement expired in late 1990. So in February 1990, after some very difficult negotiations, Williams Holdings took over Hartnell Microelectronics.

The increased resource that Williams Holdings was able to provide has made the HART detector the leading high-sensitivity smoke detector in the world. The company now employs 60 people in the manufacturing of the HART detector.

In reflecting upon the technology-transfer deal that led to the development of the HART detector, during a 1994 seminar in Newcastle, England, Keith Hartnell commented on the process [1]: "Technology transfer allowed us to get our hands on frontier technology at a relatively low cost. We could never have done that with our own resources."

From this case study we can highlight a number of important points:

- It is important to establish the nearness to market of the technology. It may require significant effort, resources, and time to get it into shape.
- A major key to success was that Hartnell did not try to create a new market for a new product. It first developed a new market for the existing VESDA product and then later introduced the new HART technology to this now existing market.
- Besides taking on the technology, Hartnell had to recruit technical specialists and establish production facilities. The organization had to be prepared and ready to receive the technology. All of this requires investment. Not only does suitable technology need to be sourced but funds must be acquired to develop it.
- The licensing agreement is very important. Eagerness to sign up a product needs to be tempered by the problems that can occur at a later date if the agree-

ment is not right. Both partners must be comfortable with the arrangement. In this case, the high royalty left Hartnell with little room to maneuver when the product component cost became a problem.

- Technology-transfer opportunities may in themselves not be suitable, but they may identify people and organizations working in areas that may be of interest. The original Hatfield Polytechnic reference was not suitable, but it did lead to RSRE and, eventually, the right technology.
- From the inception the commitment to the HART detector was prevalent throughout the company. With Managing Director Keith Hartnell acting as product champion, its progress was assured.
- It is pointless to take on new technology without a marketing plan to exploit it. Hartnell succeeded because of the company's marketing capability.

5.4.2 Example B: Cambridge Power Beams, U.K.

Cambridge Power Beams was formed as a direct consequence of a project in the EUREKA program. The aim of the project was to develop a nonvacuum electron beam (NVEB) welding machine. This would use a beam of high-speed electrons to fuse materials together. Unlike previous systems, this new version would not need to work in a vacuum; it would also enable manufacturers to weld deeper and faster than ever before and with even higher precision and flexibility. Any one company in this field could not develop the system on its own because of the range of technologies and resources required to do so. Therefore, various companies across Europe joined together to develop their own technologies for each development stage of the NVEB. In collaborating to produce the complete system, the companies shared in the risks and costs of development.

A further benefit is that the high-voltage technology developed during the project is expected to be sold through the company, and there will be spinoff effects for other products in the partner companies.

5.4.3 Example C: Renault Motor Company, France

French car manufacturer Renault is researching new materials technologies that it hopes will provide a distinct advantage in the very competitive world market. Renault is collaborating with six partners, which are all leading manufacturers of the materials suggested for testing. The partners are based in Italy, Norway, Switzerland, and the Netherlands. Renault's aims are to extend the range of body types offered to the customer, to anticipate antipollution regulations, and to satisfy customer demands for a more economical car. In addition, the company wishes to improve flexibility and reduce environmental impact without increasing the costs of capital investment or manufacturing processes. Central to the research is the selection of the best materials for each part of the car, given the criteria that strength and stiffness are important. The materials must also be light, easily recycled, and well suited to modular construction. In the first phase, the company identified two feasible options. One option is using only steel, and the other option makes significant use of aluminum and plastics. The rival technologies will be tested against each other, and prototypes will then be built. The partners are working on the innovative use of traditional materials and on the development of completely new ones. The partners claim that there are good information flows between all of them, and for the moment any partnership problems are solved very quickly even though technical ones take a little longer. Most importantly, all of the partners believe they will benefit from the project. It is foreseen that many of the technologies developed will have wider applications to other companies. These are expected to be taken up by manufacturers of products as diverse as railway carriages and electric household appliances.

Besides the high-technology work between the companies, the major points arising from examples B and C are the spinoff technologies that will result from the collaborations. The companies will learn a great deal from the research and development and should gain a competitive advantage in the marketplace by having leading-edge technology.

5.5 BARRIERS TO TRANSFERS

So what hinders the process of technology transfers between companies or causes a joint project to fail?

For technology transfer the problems are generally:

- *Lack of awareness*—many companies, particularly the smaller ones, do not know what technology is available to help them. They are often unaware of who (agents) or what (information systems) can help in the process.
- *Lack of knowledge*—technology transfer is a means of diffusing knowledge. If the staff of a company needing the new skills is lacking technical knowledge, it may not be able to use the technology being offered in the transfer.
- *Lack of funds*—the company might not be able to afford the new technology or might not be able to finance the development costs to the new product.
- *Lack of common interests*—this often leads to lack of motivation to reach agreement and a difference in opinions about the options. It can also lead to individuals putting the interests of their own company ahead of the alliance.
- *Conflict of interest*—this leads to adverse effects on the competitive advantage of the companies. Even when an excellent relationship exists, it has been found that collaborations between competing companies do not work. A take-over situation is a more likely possibility.
- *Lack of trust*—if there is little trust at the start of the venture, ways must be found to build it or the chances of successful cooperation are slim.

- *Poor communications*—it is often very difficult to keep each other informed about the activities, thought processes, goals, and direction of the venture.
- *Lack of resources*—some companies do not have the equipment to take on new technologies. A supply company, for instance, may not be able to electronically exchange data with the main contractor because it does not have an appropriate computer or the staff does not have the skills to use it.
- *Lack of time*—even though a company may need to change direction and develop a new product, it is so busy with present business that it cannot devote time to planning for growth. It has no time to research for the new development. This has proved a fatal mistake for many companies whose old products have become dated and lose market share.

With regard to collaboration on projects, there are some additional factors that can prevent success. The four main factors have been found to be:

- *Technical problems*—which generally are overcome, but which can add time, costs, and frustration to the project;
- *Resource limitations*—due to uncertainty about funding or poor budget control;
- *Problems caused by changes in the project's structure*—either through the withdrawal of partners or the loss of key members of staff;
- *Organizational problems*—due to a partner losing or changing interest in the technological area.

One of the above factors, the loss of key members of staff, requires particular emphasis. This is often overlooked in company policy. All organizations need to motivate and retain staff.

Apart from better products and generally increased market share, which are essentials in any business venture, most companies involved in forms of collaboration report other benefits. There were many unanticipated spinoffs that could be commercialized. These products also came to the marketplace earlier than otherwise expected. Working together has created better networks for the exchange of information and for improving the skills of personnel. Although there are many obstacles to successful collaboration, most companies that have experience of such report that there is a lot to gain—and very often little to lose—by working together.

5.6 CONCLUSION

Technology transfer and collaborations between companies are very important dimensions in the strategy of firms attempting technological innovation. The importance of collaboration is particularly seen in Japan, where more joint-research programs among companies in different industries are being initiated. This will link different areas of knowledge to generate new technologies. Collaboration also provides a means by which companies can observe novel technological develop-ments without having to undertake the expense and risk of investing in specula-tive research. It can also mean that a company can produce a product that it would otherwise be unable to do because it lacked the necessary technology.

As more collaborations become international, managers should be aware of national and business cultures and organization. They will also need to know about the differences in the working practices between countries. The success of technological innovations depends increasingly on the manner in which compa-nies collaborate with other companies. The success of many companies will de-pend on the level and tempo in which they develop and implement technological innovations. This is also very strongly linked to forming successful collaborations.

Finally, the success of the collaboration between companies largely depends on the degree to which the interests of the companies complement one another. There needs to be the definite opportunity for all partners to be winners.

References

[1] "Making Technology Transfer Happen," *Euro-Innovation Managers Project*, the Regional Technology Centre, Sunderland, U.K., March 1994.

Chapter 6

Evaluating the Technology

6.1 INTRODUCTION

We have now completed tasks 1 to 3 in the TT list. We are aware of the technology we require and where to find it. Therefore, the next step is to evaluate the technology.

A systematic method for the selection and evaluation of new proposals is a natural counterpart to the innovative atmosphere that produces the proposals. Criteria for the assessment of new proposals have to be established. Assessment questions include:

- Is the proposal consistent with the company strategy on quality improvement, development of new products, or reducing costs?
- How does the proposal relate to existing systems of activities or other proposals in the pipeline?
- What are the advantages, risks, and implications of the project itself?

6.2 STRATEGIC CONSIDERATIONS

Unless very good communication exists in an organization, there is a possibility that strategic priorities may be ignored when decisions are taken at lower levels. These departmental or divisional decisions may be based not on strategic objectives, but on objectives that are parochial or even driven by expectations of reward or recognition.

There are still companies in which the selection of possible new technology is driven by technological considerations rather than as part of an overall response to customer needs, market imperatives, or quality. However, equally, management may not be responsive to or understand technological factors. Typically, management is accused of being driven by short-term profit making or in love with new market ideas. Good communication systems can only be successfully established if there is genuine and sustained commitment from senior management.

Many practical things can be done to improve communications within the company: put a notice board in a particularly visible location where policy statements, company updates, and general information can be displayed; write a regular newsletter with contributions from different parts of the workforce; break down interdepartmental barriers by positioning workers close together in offices; or establish a scheme where good service from one department to another is rewarded. Whatever mechanisms are used for communication, a genuine and sustained commitment from senior management is essential.

6.3 SYSTEM CONSIDERATIONS

Any new technology process is likely to have a domino effect on other systems within the company. For example, a new flexible manufacturing system in a factory will require different quality material from suppliers and different time schedules for storage and delivery of the finished product. Also, the new process may no longer require procedures or material used by other processes (thus affecting their ordering, costing, etc.). In introducing new technology, some new processes or different working techniques around an existing system may be required. The new technology precipitates changes for which an understanding and acceptance is vital.

The overall effect of the new technology can thus be much more wide ranging than the proposer might originally think. Therefore, the proposal, while intrinsically desirable, might be rejected because of its total effect, although the possibility that the system is simply insufficiently flexible to allow change must be considered. The rejection of new technology will require careful handling to act as a useful staff development exercise on how the company works and to maintain enthusiasm in the proposer.

6.4 THE PROJECT

A formal system is essential to impose on the proposer the need for a rigorous justification. This will include financial appraisal of the kind that will be explained in Chapter 8. It will also include the need to assess risks (risk analysis will be covered in Chapter 11).

But some of the overall evaluation will have to go beyond the limitations of financial appraisal and risk assessment. For instance, each of the following effects requires careful evaluation:

- Strategic implications;
- Effect on market and on customers;
- Research and development implications;
- Operational changes in the company and for suppliers;
- Personnel;
- Training.

The evaluation system must ensure that:

- All factors that have a bearing on the total implications of the proposal are examined in detail;
- Those involved are clear about the need for a detailed justification;
- The critical factors for success (or failure) are identified;
- The need for additional information is shown;
- There is a common basis for comparing proposals;
- There is a procedure for obtaining information and commitment from all those affected by the change;
- Assumptions are recorded for future use.

Continuing evaluation of the proposal is essential after the initial agreement. This ensures both good control over the project and improves the understanding about innovation and change in the company.

Consider the following example of a product-range evaluation based on the work of the leading researcher on evaluation, D. L. Replogle [1]:

IPS, a real but disguised builder of diesel-powered heavy equipment, commissioned a study in 1988 to determine whether it needed to invest in the development of a new engine to maintain its competitive position in the 1990s and beyond. Answering this question required an analysis that correlated the needs of its customers with the technological possibilities available to the designer of a new engine.

6.4.1 The Market Analysis

The first step was to segment the market for IPS products on the basis of user needs. After an extensive review of published data and a series of interviews with end users, it was determined that all of the users fell into one of three segments (the following profiles are greatly simplified from those actually used):

- Segment A, representing about 70% of the units in use, bought the product for what amounted to virtually continuous duty. They demanded exceptional fuel efficiency, long service, and high reliability.
- Segment B, which comprised about 20% of the units, had duty cycles only about a third as severe as Segment A. The units were installed in less critical

applications than Segment A units; however, they still required reasonable reliability because backup units were rarely available.

- Segment C (10% of all units) ran very light-duty cycles in areas where backup units were always available.

The consultants generated life-cycle cost models for each of these market segments. The models considered the effects of initial price, fuel consumption, scheduled maintenance, rebuild intervals, rebuild costs, and salvage value in arriving at a life-cycle cost for any engine being used in any segment. Because IPS products served a mature market where no additional capacity was required, customers could always defer the purchase of new units by rebuilding old ones. Accordingly, rebuilding old units was termed the "baseline behavior" of all segments. Thus, the present value of any new machine configuration in any segment became the price of the product when the end user was indifferent about buying the new engine or rebuilding the old one, because the two life-cycle costs were equal.

Although the interviewed customers in all categories expressed similar wish lists of product attributes, the models reveal that each segment makes different tradeoffs between attributes. For example, a fuel-economy improvement of constant magnitude had only one-third the present value for segment B that it had for segment A, and less than one-tenth the value for segment C relative to segment A. This quickly revealed why IPS, with its high-cost fuel-efficient engines, had been so much more successful with segment A than with segments B or C.

6.4.2 The Technical Analysis

Because the market model indicated that the major determinant of customer value was the tradeoff achieved between initial cost, fuel efficiency, and service life, the technical analysis was designed to predict how each of these quantities would be affected by changes in key engine design variables. An analysis of state-of-the-art engines combined with a review of the technical literature revealed that these key variables were engine displacement, operating speed, peak operating pressure, and mean operating pressure. By analyzing thermodynamic simulations to study combinations of these variables, the team was able to map the possible combinations of initial cost, fuel efficiency, and durability that the new engine might achieve, and thus it was able to determine the customer value of each alternative.

6.4.3 The Results

The combined analysis revealed that it was feasible to create an engine for Segment A that would be worth far more to customers than IPS's current product range or that of its competitors. However, the analysis also revealed that the new engine would be no more successful in segments B and C than the current unit.

In fact, the best engine technology for B and C was so different from the one required for segment A that a common design was not feasible. This led to a final recommendation to develop a new unit for segment A, while buying units for Segments B and C from other manufacturers.

Thus, measures of utility and cost are useful only if they are sufficiently precise to guide real engineering tradeoffs. In a similar example to IPS's experience, most buyers in the "performance" segment of the car market will say they value acceleration, but the factors that limit acceleration vary across the speed range. Thus, we cannot properly respond to a demand for acceleration unless we know whether our customer wants to burn rubber when the traffic light turns green, pass trucks at cruising pace, or speed over fairly short distances. Depending on the answer, the best response may be improved traction, revised gearing, more power, or a reduction in drag.

New technologies are often undervalued because they are looked upon as straightforward substitutes for the technologies they will replace, without noting that the new technology may overcome limitations that restricted the usefulness of the old one. For example, as the transistor supplanted the vacuum tube, it overcame the thermal limits that prevented the use of a large number of tubes in a single machine (such as a computer). Later, as the integrated circuit replaced the transistor, it overcame the cost of interconnection that limited the maximum feasible quantity of transistors. As a result, both technologies showed a rate of growth that could not be predicted by extrapolating the growth of the technologies they replaced.

Also, the new technologies may have a wider applicability than those they replace. The handheld calculator is used in a much wider range of customer circumstances than were the old, larger desktop units.

We have now covered all of the aspects that form the factors on which to base a case to introduce change. So, having identified the technology for a product or process which has been evaluated and appears to have benefits for the company, it is necessary to ensure that the case for the introduction of the new technology is well presented to the company's senior management. If you are unfamiliar with presenting a written proposal, the following will help as a checklist in making sure you include the necessary facts in your report.

If the funding for the new development is to be found entirely inhouse, then the report may be used as a sort of internal business plan (business plans are covered in detail in Chapter 7). If the financial support is to be obtained externally, then the report could form a useful first stage towards the external business plan.

6.5 PRESENTING THE CASE

The report will need to be concise and address the important points. Suggested headings in the report are:

- *Background.* The background should address the following questions: Why was there a need to carry out a search for new technology? What is the evidence to support the search decision? If the search is in response to a market opportunity, how was the opportunity identified? Include any detailed evidence in an appendix.
- *SWOT analysis.* Details of the strengths, weaknesses, opportunities, and threats (SWOT) analysis carried out prior to the search should be included as the basis for the search criteria and evaluation.
- *Search criteria.* What were the search criteria regarding resources (including finance), timescales, sources of technology (HE, private research companies, government laboratories, etc.), technology bases, geographical areas, and support? The terms of reference for the search should be given in an appendix.
- *Evaluation.* What were the evaluation criteria and guidelines? How was the technology assessed for suitability?
- *The search.* Provide a brief description of the stages of the search process. Details such as organizations and persons contacted should be listed in an appendix.
- *Technology evaluated.* Depending on the number of technologies evaluated, it may be preferable to include the full list as an appendix. This section of the report should include only those that pass evaluation. The appendix should include each item, along with the criteria by which some have failed. The stage of the development should be clearly stated, and should note whether further research is needed.
- *Benefits.* State the possible benefits to the company, such as increased market share, technological ability, or engineering ability. Qualify where possible.
- *Suppliers and intermediaries.* Include a short profile of suppliers and intermediaries involved, covering such items as size, turnover, location, key personnel, product range, industry sector, R&D capability, and technological success.
- *Competition.* Who are the possible competitors, and what stage has their development reached? What is their potential *vís a vís* your company?
- *Company policy.* How does the technology fit in with the company's mission and strategy?
- *Conclusions.* This should provide a summary of the particular technologies you believe are suitable for your company to consider. Summarize your arguments supporting the cases.
- *Recommendations.* This is perhaps the most important section of the report. It should include actions you consider necessary for the company to carry out to adopt the technology you are recommending. The recommendation could be for the company to do nothing or to undertake a further search. Whatever the recommendations, they should be supported by reference to the arguments in the body of the report.
- *Executive summary.* If the report is long it should have an executive summary, which will highlight the main points, have a conclusion, and include

the recommendations. Writing an executive summary is not easy, but it is extremely important; it will probably do more than any other part of the document to influence the decision making. It can also favorably dispose the reader towards a venture at the outset, which is, of course, what you want. The following is an example of a fictitious, but illustrative, executive summary:

- Scientex will assemble and market a new design for a document shredder, which is already in prototype. The design work was carried out by the author of this report with the company's design section (using commercially available technology from Texas State University). The shredder is somewhat special in that its design, based on waterproof equipment from the food industry, is aimed at the outdoor industries market and fulfills a need so far overlooked by other manufacturers.

- The shredder offers to the customer an portable machine that is light to carry yet robust, small to store, and readily accessible. It is battery powered and has a carbonizing feature for the shredded output. It readily produces a secure environment for paperwork used on construction sites, farms, or for staff away from offices.

- It performs as well any comparable, wired machine currently on the market. The shredder can be produced at low unit cost. This, together with the market price it can command for its features, will mean that only a small turnover is needed for the company to break even. There is the prospect of a sizable market both in this country and abroad, with the chance of the company making large profits.

- The company should concentrate on this product for the first five years to ensure that it reaches all of its potential market. This will provide a base from which the company can expand into other products or incorporate the manufacturing side of the product into its own capabilities.

- An outline financial appraisal shown in the report indicates that the new development will exceed standard company net present value targets by 50% over a 10-year period. This assumes that the sales turnover will be over $2.5 million by the end of the third year with a gross profit of $0.9 million.

- We will need an investment of $0.6 million to implement our strategy, with roughly a third going into a 10-year licensing arrangement with Texas State University, another third going into tangibles such as new equipment and stock, and the rest for marketing and development expenses.

6.6 CONCLUSION

Since Chapter 2, we have been looking at ways to establish the options and opportunities for a company. Essentially we have concentrated on the external aspects that the company will need to investigate and analyze. This can be summarized by referring to the technology-transfer process on the TT list, namely tasks 1 to 4. That is:

- Recognizing a market need;
- Identifying appropriate technology;
- Securing the technology;
- Evaluating the technology.

If you have not done so already, now would be a good time to consider how you might fill in the responsibility and actions/sources columns in the TT list for tasks 1–4.)

The case has now been presented for the adoption of a new product or service. In Chapters 7 to 10 we will look at the next stages in the process. The management of the company has agreed to pursue this new venture further, so the internal analysis must take place from business plan to product development—tasks 5–10 in our process model.

References

[1]   Replogle, D. L., "Guidelines for a Product Technology Strategy," *Planning Review*, Vol. 16, No. 6, 1988, pp. 12–18.

Chapter 7

Creating a Business Plan

7.1 DECISION MAKING: THE SITUATION SO FAR

In Chapter 1 we looked at the "why" of innovation, and in Chapters 4 to 6 we examined the external factors affecting the process. The table deliniating the technology-transfer process in Chapter 2 (Table 2.1) shows that we are now at the stage where the company needs to assess the status of the internal factors influencing the decision making:

- The creation of a business plan;
- Obtaining finance;
- Establishment of market potential;
- Protection of the technology;
- Production of a prototype;
- Product development.

With respect to time, the company has looked at its options and opportunities, the necessary people have decided where the company should be going and have made their presentation of the facts to the management board, and the board has decided to move to the next stage in preparation for adopting the new technology.

At this point, to consolidate how the various criteria operate in practice, let us look again at the diagrammatic model of invention, innovation, design, and diffusion illustrated in Figure 1.1. In general terms, this "closed loop" is a good representation of the technology-transfer process stage we have now reached in Chapter 7—with no obvious starting or finishing point and a "decision maker" reviewing the whole process. The process of decision making is, as you would probably expect, complex. For instance, inventions may not be generated specifically in response to need (or even in response to economic market demand); equally, the product, although selling, may not be what is genuinely needed. But this simplified cycle does illustrate very well how the decision maker at the center must review each stage using some sort of criteria as a basis. In Chapter 6 we looked at the criteria to be applied to external factors, and for the remainder of this book we will look at the criteria to be applied to internal factors. The criteria in the internal phase are essentially commercial and technical. Does the invention work? Can it be produced economically? If so, will it sell? And, above all, will it make a profit?

Thus, for the remainder of the book, we will consider the various aspects to be assessed in the internal analysis; that is, the technical, marketing, and financial elements, as well as product development. But we must remember that assessment and evaluation is likely to take place at each stage. For instance, after the development (to prototype-stage possibility), a more detailed production-cost and marketing-estimates evaluation will be needed, followed by detailed design-for-production work, possibly with further tests, further followed by actual production, and then marketing and distribution. Subsequently, there may be an evaluation of consumer response, and the cycle may repeat with product modifications.

Because the technology transfer process cycle is relatively closed, the criteria adopted are mainly those of the decision maker and are thus usually commercial. Naturally, the criteria may gradually or rapidly be changed, perhaps in response to changes in the market place, legislation on product safety, or long-range corporate policy.

In reality, the individuals at each stage of the process may try to introduce their own criteria. For example, inventors may be motivated by a whole range of concerns—the effect of the product on others, its effect on the environment, personal considerations and ambitions, and commercial issues. Battles based on differing assessments of the most important priorities between the production, finance, and marketing departments are not uncommon. Even so, they tend to operate within a general corporate consensus and almost certainly within a consensus on criteria and overall technology policy. How can we ensure that there is a clear decision-making process to guarantee that a company will make decisions for the right reasons (i.e., commercial ones)? Unfortunately, there are no guarantees, but in the following sections we will signpost the major factors that all companies should consider to increase the likelihood that correct decisions are made.

7.2 AN INTRODUCTION TO BUSINESS PLANS

A possible indication of the importance (and fashionability) of the *business plan* as a concept is the publication of 14 books on business plans in 1994 alone! Two recommended titles are listed at the end of this chapter.

Generally, these books deal with companywide plans for new companies. However, the techniques can be used (in adapted forms) for introducing new technologies into an established company or for individual projects.

7.3   REASONS FOR USING A BUSINESS PLAN

There are several reasons for developing and following a business plan.

1. It will be required to convince potential sources of financial backing that the risks of the new development have been considered. This is especially important in that many sources of funding are still not accustomed to dealing with new technology.
2. In its early versions it will be useful for "selling" the new technology idea within the company.
3. The planning involved in producing the business plan pays great dividends in avoiding possible later mistakes (e.g., on market targets or on competitors). A surprising number of small and medium-sized companies do little formal planning for new developments. In Chapter 11 the use of *scenario planning* is described and recommended. Scenario planning can help considerably with the formulation of a business plan.
4. The availability of a proper plan reflects well on the development team and the company in general.
5. More accurate financial requirements will be identified.
6. A business plan enables subsequent development of the project to be monitored against the original objectives.

7.4   TYPICAL CONTENTS OF A BUSINESS PLAN

Most business plans contain the same basic ideas and contents.

- Description of the company, its mission statement, and its action plans to achieve the mission. Mission statements have become very common and are often very bland, but one example of an excellent technological mission and vision statement is that of the ASG Group within the Henkel Corporation [1]:

GROUP MISSION
We are dedicated to providing timely, reliable, thorough and cost-competitive physical and chemical analytical services as required to meet our customers' needs and expectations

GROUP STANDARDS FOR EXCELLENCE
Customer Service—We will satisfy our customers with full service expertise related to chemical and physical analysis by:
(1) Providing precise and accurate analysis from the Group's laboratory
(2) Providing technology transfer and consultation to Henkel Corporation

technical groups through an extensive information network with their laboratories

(3) Providing referrals from an external network with university, government and industrial laboratories

Technical Leadership—We will be recognised as experts in our disciplines by being continually aware of and utilizing state-of-the-art techniques and services

Work Environment—We recognise that to become a leader in today's competitive world market we must have a challenging, safe and stimulating work environment. We will take full advantage of the resources, partnerships, training, professional societies and career opportunities offered by Henkel. We will enhance the Corporation's reputation by publishing and presenting analytical achievements. We will encourage open (and face-to-face, when possible) communications with members of the research and development facility, our customers and our partners in other laboratories.

- Description of the staff and resources of the company, including outside advisers.
- Description of the present products/services plus a brief history or track record. A *new product audit* might be included here.
- Description of the proposed new technological development, including its proprietary position (e.g., patents or trade marks) and proof of its technical ability to do what it is supposed to do. A technology audit would be used here.
- Market research (covered in detail in Chapter 2), including market size, customers and their needs, competitors and their characteristics, and scenario planning.
- Competitive business strategy, which includes pricing, advertising/public relations (covered in Chapter 13), and distribution
- Operations planning, which includes selling changes (covered in Chapter 13), manufacturing changes, materials, effects on people, and suppliers or subcontractors.
- Financial appraisal/investment decision making for the new development (covered in Chapter 8).
- Controls and performance measures both now and in the future (covered in Chapter 15).

It is no coincidence that the last six items are major parts of other chapters of this book. They represent the distilled versions (designed to be read by "outsiders") of the planning processes required to make the introduction of the new technology a success.

7.5 CONCLUSION

There are some well-established *don'ts* about business plans:

- Don't pass on the whole exercise to your accountancy section or an outside firm of financial advisers. They may indeed be able to give invaluable help but the plan must be *your* plan. You must understand and agree with all of the conclusions reached.
- Don't engage in false optimism. Be realistic, especially on early sales predictions. Most financial investors will require forecasts at several levels of projected sales to feed into your appraisal. They will require justifications of your assumptions and targets.
- Don't, however, overplay the difficulties. Risks are inevitable in all new technology ventures, and the task of the business plan is to identify them beforehand and show how they can be minimized.
- Don't make the plan indigestible. Keep the discussion fairly short (a maximum of about 20 pages) with, if required, more details in appendixes. A clear, well-presented summary page (written after the business plan has been prepared) is essential.

References

[1]   Giordan, J. C., and A. M. Ahern, "Self-managed Teams," *Research. Technology Management*, Vol. 37, No. 3, 1994, pp. 33–37.

Selected Bibliography

Barrow et al., *The Business Plan Workbook*, 2nd Ed., London: Kogan Page, 1992.

West, A., *Business Plan*, London: Pitman, 1991.

Chapter 8

Financing Innovation

8.1 INTRODUCTION

Some developments arising from technology transfer may require only internal funding from within the company. This will still, of course, mean the need for considerable justification. Hopefully a good business plan of the kind described in Chapter 7 will help make the case clear.

Whether the development requires internal or external financial support, however, the appraisal and costing method must be thorough and able to fairly evaluate new-technology proposals.

Costing covers both the costing of the new-technology development and costing of products (which is discussed in Section 8.7). A typical checklist of costs (other than capital and maintenance costs) of a new development might be as follows:

1. *Technology:* the technological exchange, including feasibility studies;
2. *Technical documentation:* the transfer of documentation for the design and engineering processes;
3. *Technical assistance:* the adaptation, modification, and building of proto-types, as well as further R&D;
4. *Training assistance:* training for new processes for customers or suppliers;
5. *Transfer mechanism:* licensing, franchising, R&D collaboration, marketing agreements, and patents;
6. *Contracts:* consultancy fees, warranties, and guarantees.

8.2 SOURCES OF EXTERNAL FINANCE

The approach a company must take to best secure external financial support for a technology-transfer development is still very much tied to the venture capital or innovation culture of the country in which it is based.

Thus, in the U.S., there is generally good support through the Nasdaq stock exchange and venture capitalists, as well as a culture supportive of startup companies (often spun off from larger organizations) without as much risk of personal bankruptcy as in other countries. There is 2.5 times as much venture capital per capita invested in the U.S. as there is in the EU.

In Germany, the relatively small shareholding base is compensated for by an enlightened *communitarian capitalism* with banks more willing to lend over long periods.

In Britain, many innovators will still be faced by banks ignorant of the long-term requirements of technology transfer, or they will be subject to a similar "short-termist" attitude if capital is sought through shareholders. The government has sought to support innovators with the Enterprise Investment Scheme (EIS) and venture capital trusts (VCTs). In late 1995, the London Stock Exchange will launch its AIM alternative investments market for startups and new innovative developments.

The East Asian type of capitalism of the "small tigers" (often accompanied by political structures different from the West) supports innovation with government-corporate resources.

The culture of the country then provides evidence of the domino effect, such as better business plans in the U.S. compared with Europe because of the country's greater experience of startups.

Most countries (and regions/states within countries) have government support schemes for attracting inward investment and job creation. Within these support schemes are some opportunities for the support of technology transfer–led developments (e.g., grants for consultant advice or information services from regional technology centers). However, many small companies complain of the time required to both apply for and wait on decisions about obtaining quite small amounts of financial support.

An often underused source of funding for small companies is that of prizes. A good example is that of our continuing case study, Seabait. In 1986 (during its crucial and difficult early days) the partners submitted their first business plan for an "Innovation for Business Award" run by the Student Industrial Society and sponsored by Tate & Lyle and Touche-Ross. Winning the first prize generated a lot of good publicity and credibility for their innovative ideas. The business plan also won a "Small Business Competition" prize sponsored by Lloyds Bank, with further financial support and prestige. Their market research and financial planning were especially commended.

8.3 FINANCIAL APPRAISAL

The business plan will require standard basic accountancy details, such as projected sales, balance sheets, and cash flow figures. However, investing in new technology will probably require an exercise in selling to the financing body, especially if it has little experience with new technology. This section draws attention to the possible problems with this selling exercise and provides information about how to clearly explain the benefits of the proposed investment.

As we have seen in earlier chapters, many European countries (especially Britain) have a poor record of investing in advanced technology when compared with Japan or the U.S. Some senior engineers have claimed that their companies' financial appraisal procedures are biased against technological investments. Of course, rigorous and consistent procedures are needed, but the assumptions behind the procedures should be examined. A special problem in some countries comes from the reliance of companies on *payback* to evaluate all of their investments (surveys show over 50% of British companies use payback). The idea of payback—using length of time expected for reimbursement as the main criterion for lending money—is easy to understand and is tempting in an uncertain economic climate. However, it takes no account of the timing of cash flows or what happens after the end of the payback period. Also, of course, technological innovations cannot be guaranteed to yield immediate returns.

An alternative to payback is the use of discounted cash flow (DCF). There are two DCF techniques: internal rate of return (IRR) and net present value (NPV).

To the busy "innovation manager," these seemingly specialized concepts may indeed seem too specialized for attention, but an ability to understand the basis on which the decisions of financial backers are made is important.

The principles of DCF are the same as those for a savings & loan mortgage. With a mortgage, the rate of return (current interest rate) is known, and the value of repayments is what is calculated. With DCF, the value of repayments (annual savings) have been estimated and it is the rate of return that is calculated. These days, DCF can be done speedily with standard business software.

Even companies using DCF techniques experience a frequent problem of setting excessively high target discount rates (e.g., 30%). This will disadvantage products with longer production lives.

In addition to having to find unrealistically high levels of savings when using the payback method, many pioneers wishing to introduce new technology can find it difficult to quantify the benefits of the new innovation. This often arises because the savings will occur elsewhere in the company and because of the need to estimate (rather than calculate) their magnitude.

An example of concentrating on companywide rather than departmental objectives is shown by the introduction of robots. Robots are portrayed as machines that replace workers. However, there are other potentially more profitable applications for robots, such as improving product quality (with subsequent effect on sales) or when dealing with fluctuations in sales, where robots (unlike workers) can be switched off.

Similarly, computer-aided design (CAD) has often been justified by savings in the drawing office when drawing-office costs are usually less than 1% of the cost of sales. Therefore, an improvement in sales of just 1% due to CAD (from stand-ardization of components or faster introduction of new products) would outweigh all of the drawing-office savings.

In addition to a proper NPV assessment, there is an interesting rule of thumb used by American venture capitalists when approaching a company seeking to ex-pand by investment. This is called the 25% rule and is based on the premise that gross profit margin tends to express the company's underlying economics. By this rule, it is believed that companies with gross profit margins of less than 25% after the new investment seldom will accumulate major shareholder value. There are, of course, exceptions, but innovators who ignore this rule may find that they have devoted their company's efforts to working for wages without growing capital rewards.

An example of this was described at a recent seminar. A president of Euro-pean operations for a major U.S. company, an accomplished executive with a 30-year track record of exceptional achievement, wanted to act on his lifetime dream to build his own company. He took early retirement, a $2 million pension, and be-gan a 15-month search for a company to buy. This culminated in acquiring a $10 million regional company. Everyone assumed that this highly competent execu-tive with training in sales, technology transfer, manufacturing, finance, and mar-keting would transform his new company's sluggish historical performance into success. Five years later, however, the new entrepreneur sold the business for a quarter of his purchase price, narrowly escaping personal bankruptcy. He had spent a lifetime of preparation. His commitment was unflagging: 80 hours per week were routine. He used all of his skills. He invested heavily in new technology. He motivated the salesforce and installed performance incentives. Yet he failed.

What happened?

What happened was that he had bought into an intensively competitive business in a declining market, which demanded completely uncompromising quality. His initial gross profit margin was less than 20%, and this was made worse by the investment in new technology. There was simply not enough margin strength to support growth, service quality, and capital accumulation.

It is worth pondering if the horse is often more important than the rider in determining entrepreneurial success! Perhaps the skilled rider can immediately rec-ognize a worthwhile horse?

8.4 NPV/IRR TECHNIQUES

The goal of firms has traditionally been regarded as profit maximization. Profits are created when the market value of the firm's outputs is greater than the market value of its inputs. However, profit is essentially a single-period measure of company performance. By equating maximization of profit with maximum of the net present value of the firm, several problems are overcome; for example the tradeoff between short-term and long-term profits, the valuation of inflows and outflows of cash at different points in time, and the incorporation of risk into the analysis.

To calculate the present value of the firm, it is necessary to use the technique of discounting, as is used in mortgages. The value of money is time dependent ($1 now is worth more than $1 at some future date) and this needs to be taken into account when valuing the firm. A future sum of money being transformed into an appropriate discount rate allows the present values of all of the firm's forecast cash flows at various future dates to be obtained. As the cash flows are valued on a consistent basis, they may now be compared. Thus, the forecast cash flow profiles of firms may be evaluated in terms of their present values to determine the value of the firm. The discount rate chosen to value a firm depends on the riskiness of its forecast cash flows; those subject to considerable variability (risk) need to be discounted at a higher rate than less risky cash flows.

A firm can be regarded as a collection of individual projects, so the present value of the forecast flow profile relating to a project determines the profitability of the project and hence the contribution of the project to the profitability of the company. The chosen discount rate (or cost of capital) must reflect the riskiness of the individual project.

Decisions regarding investment in projects and products fall within the field of capital budgeting, which is the art of investing in assets that are worth more than they cost. The objective of the firm is the maximization of shareholder wealth, but an investment is only wealth-creating if the present value of all forecast cash inflows (from customers) associated with the project exceeds the present value of all forecast cash outflows (to suppliers of labor, goods, etc.) relating to the project. Investment decisions usually involve an initial capital expenditure followed by a stream of cash receipts and disbursements in subsequent periods. An NPV formula may be used to evaluate the desirability of investment opportunities. All major spreadsheet software packages have an @NPV calculation function.

Thus, the NPV of a project is obtained by summing the forecast net cash flow over the project's life, discounted at a rate that reflects the cost of a loan of equivalent risk on the capital market, and deducting the initial investment outlay. Hence, an investment is wealth-creating if its NPV is positive.

For example, suppose that a project requires an initial investment (I) of $1,000 and that the forecast net cash inflows (x) are $620 receivable after one year and $580 receivable after two years. The firm operates in a low-risk industry and divides its various projects into three categories: class A, where the risk is below average and the required rate of return, K, is 10% (or 0.1 as a fraction); class B, where the risk is average for the industry and the required rate of return is 13%; class C, where the risk is above average and the required rate of return is 16%. The project may be evaluated from the NPV viewpoint by using the *@NPV* software function.

- If the project falls in class A, NPV = $40, which implies the project is acceptable;
- If the project falls in class B, NPV = $0, which implies the project is barely acceptable and with no margin for error;
- If the project falls in class C, NPV = –$30, which implies that the project is unacceptable.

Investors do not like risk, and the greater the riskiness of returns on an investment, the greater will be the return expected by investors. Accordingly, there is a tradeoff between market risk and expected return that must be reflected in the required rates of return on investment opportunities.

8.5 PRACTICAL PROJECT AND PRODUCT DECISIONS

The net present value rule offers a theoretically correct financial approach as to whether an investment opportunity should be accepted or rejected, but in practice many other techniques are employed in investment appraisal (as well, of course, as broader approaches like risk assessment). An alternative to NPV is internal rate of return (IRR).

The IRR, or yield, on a project is the rate of return that equates the present value of anticipated net cash flows with the initial outlay. To calculate the internal rate of return, it is necessary to solve a formula similar to the NPV calculation formula. Thus, *r* is the rate of return which gives a zero NPV. A project is acceptable if its yield or internal rate of return is greater than the required rate of return on the project (*K*). This method of project appraisal usually gives the same accept/reject decision as NPV. Projects with positive NPVs will have values of *r* greater than *K*. In the example used to illustrate the NPV rule, the internal rate of return on the project is 13%. Hence, in our previous example, if the project is in class A then it is acceptable as the yield of 13% is greater than the required rate of 10%.

Managers are recommended to use the NPV method to assess capital projects because it is easier to handle than IRR. In addition, there are several problems associated with internal rate of return. For example, the size of a project needs to be kept in mind. Suppose there are two mutually exclusive investment opportunities—project 1 involving a $20,000 initial outlay and offering an internal rate of return of 30%, and project 2 involving a $10,000 initial outlay and offering an internal rate of return of 40%. In this case, project 1 may well have a higher NPV than project 2. Thus, the smaller project has a higher IRR and the larger project a higher NPV. If an investor has $20,000 to invest, then he can either select project 1 or, alternatively, invest $10,000 in project 2 and $10,000 elsewhere. The NPVs of the two $20,000 investment schedules may then be compared and the one with the higher NPV selected. NPV is a direct measure of wealth creation and therefore is the relevant criterion upon which investment schedule acceptance or rejection should be based.

A second problem associated with internal rate of return is that multiple solutions may exist to the IRR equation. Multiple solutions occur when the net cash flow changes sign. For example, a negative initial outlay may be followed by a positive net cash flow at the end of the first year, a negative flow at the end of the second year, and a positive flow at the end of the third year. When several internal rates of return satisfy the IRR equation, it is difficult to use this method to assess projects, whereas the NPV rule presents no such problems.

Other complications occur with the IRR rule when it is necessary to make decisions between mutually exclusive projects with different lives or with different patterns of cash flows.

8.6 QUANTIFICATION OF INTANGIBLE BENEFITS

In NPV an important aspect of uncertainty concerns the actual calculation or quantification of the base values of the variables themselves. In many cases, such quantification may be extremely difficult, in which case the term *intangible* may be used.

Intangible benefits may be considered to be of two types:

- They may be a type of benefit that is inherently difficult to quantify (e.g., the effect on company image of an improved working environment).
- They may be a combination of specific benefits, with the quantification difficulty lying in the effect of the combination. As an example, the installation of new stand-alone equipment may actually increase work in progress (WIP) inventory due to difficulties in achieving a satisfactory flow between the equipment and the adjacent operations; however, one of the synergy effects of an integrated system may well be a reduction of total WIP because of the improved compatibility of all or several operations.

The importance of intangible benefits is that they tend to embody the strategic-level savings that technology-based solutions provide but which economic justification methods ignore.

The relevant literature makes it clear that organizations are frequently daunted by the task of assessing this type of benefit. The result is that important savings are ignored and investment decisions are either deferred or taken "blind." Two common arguments against conducting technology financial appraisals often put forward by equipment suppliers are:

- Such appraisals are irrelevant as the investment should be viewed as an "act of faith" in the technology.
- Such investments cannot be expected to pass the traditional detailed financial appraisal.

The consequences of not including total benefits (and those consequences are surely relevant to all new-technology areas) are:

- Investment will not take place for applications that would prove highly profitable.
- A "fashionable" solution will be bought where a more cost-effective method should be adopted.
- The decisions as to type and sophistication of solution may be incorrect.

Four valid reasons for evaluating all technology proposals are:

- To allow a decision about which technology strategies to pursue;
- To allow objective decision making based on profitability, cash flow, and risk;
- To allow *cost monitoring* (post-project appraisal) to take place (i.e., to prevent the setting of oversoft targets);
- To concentrate on the achievement being attempted and how success will be measured rather than the pursuit of technology for its own sake.

What, then, specifically constitute intangible benefits? A typical list (based on work by Catton [1]) is:

Design and Marketing

Increased sales by:

- Reduced quotation and design lead times;
- Shorter and more reliable overall lead times;
- Improved "image";
- Ability to design and manufacture new product range;
- Improved-quality product.

Production

Reduced costs by:

- Component and tooling standardization;
- Scrap-rate reduction;

- Space and asset savings by reduced WIP;
- Less plant down time due to planned maintenance;
- Reducing stock outs;
- Better supervision utilization through less routine administration.

Finance

Reduced working capital by reduction in:

- Stock;
- WIP;
- Purchase pricing through vendor appraisal and procurement system;
- Creditor control through improved SOP system.

Quality

Reduced costs and improved service by:

- Higher levels of quality control (QC);
- Faster response to QC queries.

General (Applicable to All Areas)

Improved management time utilization through better information systems.
    Other benefits in particular application areas include:

CAD

- Standardized components;
- Faster introduction of new products;
- Elimination of unprofitable orders.

Robots

- Replacement of human labor for hazardous tasks;
- Reduced scrap/rework;
- Seasonal demand applicability (i.e., robots can be switched off).

MRP/MRP II

- Effect on sales;
- Reduced stocks of finished components and finished product.

FMS

- Output increase by elimination of machine/operator interface;
- Improved delivery/reliability;
- Improved efficiency due to optimization of batch quantity, set-up-per-piece time ratio, and component complexity.

The objective here is not to provide a complete list of intangible benefits, rather it is to illustrate their nature by considering examples. Clearly, there is no one comprehensive list—benefits will vary according to the technology acquired and the existing operational infrastructure.

The most important categories, however, could be summarized as improved quality, inventory control, customer lead times, machine utilization and efficiency, staff efficiency and morale, and customer image. Improved morale and customer image are clearly inherently nebulous in nature. Specific operating improvements, such as batch-quantity optimization or scrap/rework reductions, may be easily quantifiable in the case of the stand-alone machine. However, as synergy develops, from improved integration and rationalized working practices, these benefits may be much more difficult to assess.

So how are these benefits to be quantified? Once again, the correct approach is required before detailed measurement is attempted. The incorrect approach, which is unfortunately widespread, is to assume that the solution is a fixed entity and to estimate the improvements that it confers. A sounder approach is to work out in some detail the target improvements required by the system. This should be done initially at the strategic level, then in terms of specific operational improvements. The solution must then be tailored to provide the maximum overall achievement of these targets, and the extent to which such achievement is feasible can then be estimated. The benefits of this approach are:

- Higher savings will be driven through due to their nature as a self-fulfilling prophecy.
- The solution is built up from a problem analysis, not sold as a new concept.
- The assessment of achievability targets will be inherent in the solution design and will involve a dialogue between designer and end user with better and more accurately predicted savings as a consequence.

Three major factors are:

- The required benefits of the new technology must be identified prior to selection of the solution.
- Benefits must be tailored to the individual organization.
- Benefits must be considered from a companywide, not a localized, perspective.

Researchers illustrate the companywide nature of target savings and criticize the traditional approach to savings, which focuses on labor and variable overheads or, in the case of inventory, WIP only. Many new-technology solutions will have an effect on total stock. This is because the improved flexibility achieved by these solutions enables customer delivery dates that were previously only achievable by initially manufacturing for stock a standard product that then underwent final customization. Furthermore, because quality and delivery are improved, total sales will be increased. In a situation where WIP was typically 15% of total stock, and labor and variable overhead was only 20% of total sales, the broader view of benefits (i.e., reduced total stock and increased sales) even when viewed cautiously gives a dramatically improved result over the simple consideration of WIP direct cost reduction.

There are, however, major problems associated with providing a detailed account of target savings.

- It would prove to be practically impossible to achieve an integrated CIM solution with operating results predictable to this level of accuracy due to the unprecedented synergy effect.
- It would be extremely difficult to extract such detailed targets from managers who know that they would be measured against them.
- It is extremely unlikely that measurement systems would be available immediately to provide such accurate performance feedback.
- Actual performance will not equate to these targets for some years as the effects of running down existing machinery and working up new machinery (including debugging software) must be taken into account; underachievement may in fact be ascribable to valid upgrading problems for an indeterminate period.

A more sensible eight-stage process might be:

1. Identify problems without attempting to quantify them (i.e., sales are being lost due to a poor delivery performance).
2. Establish which problem can be reduced by an evaluation of the project.
3. Establish the project design that will reduce or eliminate these problems.
4. Estimate the cost of installing and operating the project.
5. Estimate the magnitude of each improvement (i.e., delivery time will improve from 14 weeks [±3 weeks] to 10 weeks [±1 week]).
6. Quantify these improvements in terms of company performance (i.e., sales will improve by between 5% and 15%).
7. Requantify these improvements into the format for evaluation (i.e., increased contribution to overhead recovery).
8. Perform the financial evaluation.

Here, the problem definition precedes the solution, after which achievable savings are identified. This process may be reasonably detailed and far reaching, but the conversion into immediate savings is more generalized.

Quantification involves fragmenting the intangible improvements into ultimate performance results to which it is then possible to apply global quantification. This process can be illustrated by the following fictitious conversation:

"Implementing a CAD system will allow design lead times to be halved" (intangible benefit).
"So what?"
"So we would have time to order material instead of holding it in stock" (i.e., tangible benefit: stock reduction).
"So we would deliver on time instead of late" (i.e., tangible benefit: invoiced and paid earlier—fall in working capital).
"So we would receive more orders because we would be competitive on delivery" (tangible benefit: more sales).

The quantification procedure is simple and straightforward in theory; however, obtaining estimates in practice is usually extremely difficult. The following suggestions are worth considering in any situation:

- Questions aimed at the marketing department should be framed to fit as closely as possible with available market research information.
- Executives are often more willing to quantify their problems rather than their opportunities; it may, therefore, be productive to ask, "What would be the effect on market share if delivery performance were to fall by 5%?" and then make the progression to estimation of effect of improved performance.
- Accurate assessments may be available from individuals formally unconnected with the functional area.
- It should be explained that a global estimate is required at first. Estimates are likely to be offered in the form of a range (e.g., "improved delivery from six to five weeks will improve sales between 5% and 7%").
- The Pareto effect is useful here: 80% of the benefits may be covered by 20% of the answers, so detailed estimation may be restricted to those areas.

Two important points should be re-emphasized here:

1. Estimates are only meaningful when supported by post-completion monitoring.
2. Suitable (i.e., long-term) performance measurements will be required to measure implementation success.

8.7 PRODUCT COSTING

Studies by cost-management accountants such as Jeans and Morrow [2] have established the need to see product costing as a separate management accounting objective for the purposes of profitability analysis and possibly as a basis for setting selling prices for the product/service arising from technology transfer.

The traditional approach to product costing (direct materials, direct labor, and factory overhead via a burden rate) is seen as inappropriate for a new-technology situation. The reasons are related to the current industrial climate of increased global competition, with focus and depth in narrow product areas:

- The factory floor is no longer a suitable boundary for the overhead, which should be accounted for in product costs. Jeans and Morrow argue that the boundary should be extended to the increasing number of necessary support departments/teams.
- Direct labor will not be the cost driver for many direct factory overheads, although traditional accounting practices in many companies may still do this. Activity-based costing (ABC) has developed as a method to attribute costs to activities or clusters of similar activities. Thus, non-value-added activities can be identified and minimized.
- The importance of the cost of extra flexibility (which arises from many technology transfers) should be recognized. Competition may force the need for small product-run sizes, but the costs of this must be evaluated and subsequently monitored.

Cost drivers need to be in sympathy with the strategic objectives of the business [2].

8.8 CONCLUSION

Finance, not surprisingly, is a major influence through the whole innovation and technology-transfer process. Before investment is obtained, the proposed idea must have its costs and benefits thoroughly analyzed. This may, however, take place against a background of a lack of understanding of the intangible benefits of new-technology developments and unrealistic payback-period expectations. When the scheme is launched, adequate financial monitoring (among other monitoring) must take place in a serious manner to ensure that targets are met, and the true sources of cost must be identified and controlled.

A basic understanding of modern financial techniques is yet another skill that a champion of innovation and technology transfer should acquire.

References

[1]   Catton, P. G., "Advanced Manufacturing Technology—Evaluating the Intangible Benefits," *Institute of Mechanical Engineers Conference Publications*, September 1986, p. 18.

[2]   Jeans, M., and M. Morrow, "Product Costing," *Management Accounting*, April 1989, pp. 29–30.

Chapter 9

Protecting the Technology

9.1 INTRODUCTION

Intellectual property rights—patents, trademarks, copyrights, confidential information and know-how, licensing, and franchising—play an important role in the innovation process. Intellectual property rights allow innovators to protect their new ideas or brand names. These rights protect the owner from exploitation. Giving legal standing to the rights permits people to buy and sell innovations similar to the way they buy and sell goods. The various rights are limited in their scope, require specific steps to be taken to acquire and deal in them, and bestow different rights to their beneficiaries. Entire books are written on various aspects on intellectual property rights. So, in this chapter, we can only give a broad outline of the different procedures.

9.2 PATENTS

A patent is a monopoly right to the exclusive use of an invention. In most countries a patent lasts for a maximum period of 20 years dating from when the application for the patent was filed. However, in extreme cases, 10 years of this 20-year period may go by before examination is concluded in the patent office. In the U.S. and Canada the period begins when the patent is granted and lasts for 17 years.

The European Patent Convention (EPC) of 1973 (see Section 9.3.1), which has served as a model for legislation for many countries in efforts to harmonize national patent laws, states that to be patentable, an invention must:

- Be new;
- Involve an inventive step;

- Be capable of industrial application;
- Not be excluded.

Being *new* means that the invention must never have been made public in any way, anywhere in the world, before the date on which an application for a patent is filed.

An invention has an *inventive* step if that step, when compared with what is already known, would not be directly obvious to someone with a good knowledge and experience of the subject.

Industrial application means that the invention must be a practical form of an apparatus or device, a product (new material or substance), or an industrial process or method of operation.

With regard to being *excluded* when defining what constitutes an invention, it is usual to distinguish between an invention and a discovery. However the term *discovery* is not standardized universally. Whereas the patent laws of most countries explicitly exclude *discoveries* from patent protection, the laws of other countries use the words *invention* and *discovery* synonymously or provide special regulation for *discoveries*. The EPC declares that an invention is not patentable if it is a discovery; a scientific theory; a mathematical method; an aesthetic creation such as literary, dramatic, or artistic work; a scheme or method for performing a mental act, playing a game, or doing business; the presentation of information; or a computer program [1]. If the invention involves more than abstract aspects, so that it has practical features, then it may be patentable. It is not possible to obtain a patent for an invention if the publication or exploitation of it would generally be expected to encourage offensive, immoral, or antisocial behavior.

When an invention is made it belongs to the inventor, unless the rights have been given to someone else by an agreement or contract of employment. An employee making an invention in the course of normal duties will generally not have the right of invention; this will probably belong to the employer.

Patenting is obviously complex, and the results of doing it incorrectly can be very serious. Thus, it can be beneficial and possibly necessary to use a registered patent agent.

9.3 PATENT RIGHTS

A patent gives the proprietor alone the right to the manufacture, use, importation, or sale within the country of the patent of the invention for which it has been granted.

The right given by the patent is not enforced automatically; it is the responsibility of the owner to look after the rights by finding out whether someone is *infringing* on the patent.

According to past and present understanding of patent protection, the inventor or his or her legal successor is accorded an exclusionary right. This requires that the inventor discloses the invention to the public within the patent application. This disclosure is meant to perform three functions:

- To prevent any person from inadvertently infringing on the patent right during the term of monopoly;
- To put the public in possession of the patentee's secret on the expiration of the patent term;
- To prevent patents from being granted for known inventions.

In the early days of patenting, the requirement of full disclosure was relatively easy to meet and applicants were able to describe their invention in the space of a few pages. An adequate disclosure today often requires over 100 pages and in extreme cases over 1,000 pages to describe. Hence the needs nowadays for the involvement of experts in drawing up patent applications.

In most countries the law adopts the condition of so called *absolute novelty*, which is that the invention claimed must not already be available to others by any kind of public disclosure or use before the date of filing of the patent application. The rule of novelty is generally expressed in terms of *publication* and this expression includes all forms of public disclosure and is not limited to literature publications. Thus, prior written or oral disclosure of an invention, or any other means of making the knowledge available in a public manner, before seeking patent protection contributes the invention to the *state of the art*. In the EPC the state of the art is defined as [2]: "Everything made available to the public by means of a written or oral description, by use, or in any other way, before the date of filing of the European patent application." EPC law also includes within the state of the art the contents of prior filed European patent applications that proceed to publication; these are deemed part of the state of the art from their filing date.

It follows that an inventor's own public disclosure of a discovery prior to filing a patent application makes it part of the state of the art according to the law of the countries with absolute novelty laws. Therefore, if the disclosure contains sufficient detail to enable the invention to be understood and used, it can no longer be patented.

Once the patent application has been filed, the examiner in the patent office will check its validity against the requirement of the law. The examiner will then conduct a search for patents or other state of the art that might suggest the obviousness of the claimed invention or actually anticipate it. After the search the examiner will determine whether to let it proceed or to reject the whole application or any part of it. In addition, the examiner may advise on some form of limitation of the claim

Patents and infringement of such is without doubt a complex web of articles of law. To complicate matters further, practice can vary from country to country despite attempts to be more universal in application. We have already mentioned that the period of a patent is different in the U.S. and Canada from most other countries. Similarly, interpretation of disclosure and even what constitutes a patent can be different from state to state. In fact, obtaining foreign patent protection in countries other than one's own, particularly in non-English-speaking countries, has generally been an expensive business. In almost all foreign countries, local legal representatives or agents need to be employed so that the requirements of the laws of each country are met. In many countries, annual fees are charged to maintain the patent and there can be requirements that an invention be practiced within the countries to remain active. Therefore, maintaining patents in many foreign countries can be unjustifiably expensive.

Thus, when considering protection abroad, it is advisable to look closely at the potential markets in foreign countries. If you do not intend to actively market in another county or do not know of a third party to whom you could sell or license the intellectual property rights, it may be better not to bother with obtaining protection in that country. However, if protection is not obtained abroad, others can exploit the product in those countries lacking protection, and intellectual property rights, where they exist, can be infringed upon by importation from a country where no protection has been sought. Basically, anyone could market or produce your product in a country that does not recognize your patent. However, even in a country that recognizes the patent, your product could be imported from a different organization if that organization was in a country that again did not recognize your patent.

On a more optimistic front, the broad principles of protection are recognized nearly everywhere, and although the particular provisions may vary, international agreements oblige countries to open their intellectual property systems to nationals of other member states.

9.3.1 European Route

The EPC is a convention separate from the European Union, although at the time of writing most members of the convention are or will shortly be full members of the EU. The EPC has been signed by representatives from Austria, Belgium, France, Germany, Italy, Liechtenstein, Luxembourg, Monaco, the Netherlands, Portugal, Spain, Sweden, Switzerland, and the U.K. The main purpose of the EPC is to allow patent rights to be obtained in any one or more of the countries forming the convention, using a single European patent application. The applicant has to state in which countries the patent is required and pay a fee for each. This is cheaper than making separate national patent applications in each country. The crucial aspects in European law are that the patent is granted to the person filing the application, which may or may not be the inventor, and discoveries are not regarded as inventions. European patent applications are subject to a dual publication system. A novelty search report on the state of the art is provided by the International Patent Institute at The Hague. The first publication of the patent application to seek views on patentability takes place within 18 months from the filing date. The second publication is made upon the grant of the European patent. The term of the patent is for 20 years from the date of filing of the application.

9.3.2 International Route

A national of, or resident in, one of the Patent Co-operation Treaty (PCT) countries may file an application to obtain patents in several of the countries that have agreed the treaty. PCT member states include most of the countries in the world, including such major trading ones as Australia, Brazil, Canada, Japan, Hungary, Norway, Poland, Romania, the Russian Federation, the Ukraine, and the U.S., plus the EPC member states and some countries in Africa. The PCT is administered by the World Intellectual Property Organisation. A single international application (in one language and in accordance with one set of rules on formal preliminaries), a prior art search, and publication of the application is filed at a single receiving office. The single application is sent to an international searching authority, where a novelty search is carried out. Publication, as with the EPC, comes within 18 months of the date of the first filing of the application.

An advantage of both EPC and PCT filing is that the required applications can be prepared in exactly the same format. Their form and content will be accepted in all countries that have adhered to the programs of the respective conventions. The EPC and PCT are working toward the development of a multinational patent system. However, many people believe that this is neither achievable nor desired and that many countries prefer to maintain their own patent systems.

A major disadvantage of both of these international patent filings is the cost. Before savings over a country-by-country approach are achieved, filing must be anticipated in several countries—perhaps as many as four to six, depending on which countries are selected. Other disadvantages of the EPC filing is that a single examination takes place for all of the designated countries, and patent protection in all of these countries is determined through this single examination procedure. Therefore, if an individual is unable to obtain patent protection in one member country, he or she cannot obtain protection in another.

9.3.3 Differences in U.S. Patent Laws

We have outlined some of the principles of patents, and we have mentioned that there are differences from country to country in interpretation. These differences may be in definitions of terminology or in applications of the law. Let us look at some of these differences.

In the U.S., it is the inventor who is granted the patent, rather than the person filing the patent application as in most other countries.

The words *invention* and *discovery* tend to be used synonymously in the law of the U.S. Even so, as long ago as 1862, the courts clarified the difference between patentable and nonpatentable discoveries by stating that laws of nature, though they may be discovered, may not be patented as an invention; a discovery may be patented only if the new force or principle brought to light is embodied and set to work and only in connection or combination with the means by which, or the medium through which, it operates. So the U.S. patent laws do not exclude discoveries—but, in the common broad meaning of the term, they are not within patentable subject matter.

The question of *industrial application*, which is used as a requirement in patent law in most countries in Europe and Japan and which can be difficult to define, does not arise in the U.S. The U.S. Patent Act of 1952 expressly itemizes the subject matter eligible for patent protection as [3]: "any new and useful process, machine, manufacture, or composition of matter, or any useful improvement thereof." An invention must be capable of achieving some minimal useful purpose that is not illegal, immoral, or contrary to public policy. This is easy to interpret for inanimate objects. However, interpretation of EPC and U.S. standards is more difficult in the more recent patent world of plants. This is because a true *product of nature* does not constitute a machine, composition of matter or manufacture, and may not be an industrial application. Most countries have determined that biotechnology inventions must conform to the requirements of the patent law without adjustment. It appears that the wording of the U.S. patent law is more beneficial, though, to the biotechnology industry because it does not seem to have encountered any unique problems.

We have already stated in Section 9.3 that if the disclosure of an invention contains sufficient detail to enable the invention to be understood and utilized, it can no longer be patented. However, the patent laws of the U.S. are an exception to the generality of this rule because they allow a one-year period of grace for the filing of a patent application after publication by the inventor. The laws of Japan allow a six-month period of grace for disclosures at scientific meetings approved by the president of the patent office.

Experience has shown that patenting is one of the most effective and the most secure ways of transferring technology from university to industry and commerce. Both the understanding and the use of the patent system by university researchers is probably better in the U.S. than it is anywhere else in the world. This is mainly due to the long history of entrepreneurial attitudes of U.S. researchers and universities but is also due to the patent laws themselves. The grace period works very well in the U.S., and researchers do not feel as inclined to be entrepreneurial in those countries with no or some restrictive periods. The U.S. universities have developed very successful inhouse patent administrative structures, maintaining their own patent offices with full-time legal experts. Research units in other countries are beginning to copy this model, albeit on a smaller scale with either one full-time person responsible for intellectual properties or someone who has a partial responsibility and works with a local patent agent. It is quite clear that one essential ingredient for successful technology transfer is a positive well-informed attitude from the academic sector toward patenting at all levels, from undergraduate junior researcher to senior professor.

9.4 TRADE SECRETS

It is not necessary to patent your invention. Rather than patent an invention, a person may prefer to keep the process secret and exploit it alone, or in conjunction with others under so-called *know-how*, *confidentiality*, or *noncompetition* agreements. This may be because obtaining a patent requires the disclosure of the invention. The hull design in the America's Cup example cited in Chapter 4, where the competitors tried all means to see the revolutionary new design, is a case in which the designers perhaps should not have gone for a patent immediately, but kept their design secret. (Fortunately for the designers, it seems the competitors had not thought to check for a patent, which would have been the obvious way to look at the hull.) However, there are obvious inherent dangers in not patenting, as an inventor could fall victim to industrial espionage, lose his or her lead in the marketplace, and have no legal claim on a design.

A *trade secret* may constitute any information that is not generally known in the trade. It does not have to be patentable to merit protection. Neither does it need to be new or satisfy any of the other parameters of a patentable invention.

At the time a patent application is filed, the invention it describes may represent mainly the product of research and experimentation. Once the product is in development, a great deal of extra information that is important to the actual practice of the invention in a production phase may be raised. This new expertise may take on a value equal to if not more than that of the patented object. Therefore, if this expertise is properly protected as a trade secret, competitive advantage is maintained over business rivals—perhaps to an extent where others are willing to pay high fees in royalties for permission to use the trade secret. One major advantage of a trade secret is that the owner does not lose his or her rights at the end of 17 or 20 years, as with a patent. The disadvantage is that competitive edge gained by a trade secret can disappear overnight if the secret becomes available to the public or if another researcher independently develops the same process.

Trade secrets are generally protected by common law rather than national or international agreements and articles or statutory law. They are particularly protected by contract law and the laws governing unfair competition. Because a trade secret must, by definition, remain a secret to be protectable, steps need to be taken by the owner to secure secrecy.

9.5 CONFIDENTIAL INFORMATION AND KNOW-HOW

Trade secrets, confidential information, and know-how are often vital assets to a business. The legal means of maintaining confidentiality are relatively straightforward and inexpensive. You must ensure that anyone with access to the information or to whom you propose to disclose it enters into a written agreement to keep the information confidential. In the case of employees, even if secrecy is not expressly stated in the employees' contracts of employment, it will normally be implied. However the obligations of an ex-employee are much less stringent. When employees do leave, it may be wise to have them sign a declaration, *a nondisclosure document*, which states that they have not retained any documents or other company property and that they will not disclose any information regarding present or future products or developments.

The situation is easier when dealing with clients, for then you can control both the extent and timing of any disclosure. The use of a simple secrecy agreement will suffice when negotiating the sale or use of license of intellectual property. Employees, sales representatives, consultants, and visitors should all be bound to obligations of confidentiality if you wish to keep aspects of your business secret.

Secret know-how, a form of intellectual property, generally relates to a technology process or method that is not patentable, either because it did not satisfy the criteria for patentability or because there has been a strategic business decision to maintain it as confidential or secret information. The know-how is usually exploited by transfer to the client under strict terms of confidentiality of the technology and information. The transfer usually involves a period of instruction and training for the users of the new technology. The company and those of its employees who come into direct contact with the know-how are all bound by the terms of confidentiality. Know-how is licensed under the usual terms that exist for patents and other forms of intellectual property, including license fees, deposit payments, and royalties. It can be licensed either exclusively or nonexclusively. Know-how license agreements form an important part of technology-transfer agreements. It is widely accepted that a large proportion of technology-transfer agreements relate to some form of know-how, either alone or in association with patented technology.

It is very important that all information related to know-how is exchanged only after a suitable confidentiality agreement is signed by all parties. All information provided to the client would be marked as confidential and appropriately referenced in the governing confidentiality agreement. Employees and other third parties (such as consultants and associates) who would have access to the information would also be bound by the agreement. An acknowledgment of receipt of the confidential information is useful in the event of a possible breach of confidentiality.

Any prior disclosure of the know-how to the public domain, in publications or otherwise, destroys the commercial value of the know-how. In evaluating know-how, it is essential that the same knowledge is not already freely available to the public.

9.6 ANTICOMPETITIVE LAWS

In a competitive market, companies can be expected to adopt policies that they hope will give them a competitive advantage. We have already discussed how a patent can be one means of getting this advantage. It is, however, also possible to adopt policies that give a company an advantage that has very little to do with efficiency or the objective to provide better goods or services. Product distribution by a company to select retailers only or temporary price reductions to prevent competitors from entering the market can restrict or prevent competition.

To stop such practices, countries have anticompetitive or antitrust laws. These laws are concerned with maintaining competition with the advantages of market pricing, readiness for new entries, and efficiency. An anticompetitive practice can be defined as any which has, is intended to have, or is likely to have the effect of restricting, distorting, or preventing competition. It is not the actual practice itself but its effect on competition that is the underlying principle. Practices that may be acceptable in one market where competition is strong may be unacceptable in another where there is less competitive activity. A company or group of companies that has the major share of the market, possession of a leading brand name, or intellectual property rights has the potential to affect the market to their advantage. In a monopoly situation this can cause excessive prices, reductions in the level of service, and unfair restrictions on entry to the market. To stop or discourage competitors from entering a market, a company may price goods at an artificially low figure (which is loss making for the company) or may attempt to enforce a minimum price at which their goods can be resold. Clearly, both of these practices are anticompetitive.

Generally the legislation provides for a case-by-case examination. If the practice is found to be, or likely to be, against the public interest then it can be prohibited.

Two block exemptions relevant to the area of technology transfer are currently in force. These are:

- The patent-licensing block exemption;
- The know-how licensing block exemption.

The first may cover sole or exclusive patent licenses with or without ancillary know-how. The second requires there to be substantial, secret, and recorded know-how. However, there may also be patented technology, whether ancillary or not. The know-how regulation does not cover pure patent licenses where know-how remains a secret. When there are both patented technology and secret know-how, usually the know-how regulation will apply.

There is a proposal in European patent law to replace the patent-licensing block exemption and the know-how licensing block exemption with a new "technology transfer block exemption regulation." This new block exemption is intended to extend to *pure* patent licenses, *pure* know-how licenses and *mixed* licenses involving both patents and know-how. In addition, the new block exemption is intended is to extend to mixed agreements containing ancillary provisions related to intellectual property rights other than patents (particularly trademarks, copyrights, and design rights). It may be some time before the new block exemptions become a matter of routine legal practice, though. Generally, it is prudent to engage the services of a competition lawyer to advise on the law regarding the scope of licensing agreements.

9.7 COPYRIGHTS

A copyright gives rights to the creators of:

- Original literary, dramatic, musical, and artistic works;
- Published editions of works, sound recordings, films, videos, broadcasts ,and cable programs;
- Computer programs.

These rights allow the creators to control exploitation and cover copying, adapting, publishing, performing, and broadcasting.

Most countries have no registration systems for copyrights; these rights come into operation automatically. For instance, copyright protection in the U.K. lasts for 50 years after the death of the person who holds the copyright or 50 years after the publication of the copyright if this is later.

So, how can a copyright be obtained? Some countries, such as the U.S., require a work to be marked with an international symbol (©) followed by the name of the copyright owner and year of publication. This is not essential in the U.K. However, marking in this way may assist in infringement proceedings in any country, including the U.K. Copyright is infringed upon only if more than a *substantial* part of the work is reproduced (i.e., issued for sale to the public) without permission. Because there is no formal registration of copyright, the question of infringement usually has to be settled in a court of law, which is expensive.

"Who owns the copyright?" The general rule is that the first owner in a literary, dramatic, musical, or artistic work is the author. The copyright in films, sound recordings, broadcasts, and published editions belongs to the film or record producer, broadcaster, or publisher. The main exception, and this is important in the context of this module, is when such work is made in the course of employment, in which case the employer owns the copyright.

Copyright material is generally, but not invariably, protected outside the country of origin. Most countries are members of at least one international convention in the copyright field. The most notable conventions are the Berne Convention for the protection of literary and artistic works, and the Universal Copyright Convention (UCC). Copyrighted material created by a national or resident of one country is protected in each member country of the convention by the national law of that country.

9.8 TRADEMARKS AND SERVICE MARKS

A trademark is the symbol by which the goods of a particular manufacturer or trader can be identified. Trademarks can be a word, a signature, a monogram, a picture, a logo, or a combination of these.

To qualify for registration the mark must be distinctive, must not be deceptive, and must not be capable of confusion with marks already registered. Trademarks and service marks provide protection for the goodwill and reputation of a company in its products and services. So, for example, while the formulation of a washing powder sold under a particular trademark may change many times over the years, a trademark ensures that only the company or its licensee may sell a washing powder under that name.

To register a trademark you or your agent conduct preliminary searches at the trademarks branch of the patent office to check there are no conflicting marks already in existence. You then apply for registration on the official trademark form and pay the appropriate fee.

It is advisable to register a trademark, if possible, because to succeed in a court action based on an unregistered mark, it is necessary to show that one has established a reputation in the mark and that the use on which the complaint is based would be liable to confuse or deceive the public.

An eye-catching or distinctive trademark can be an integral part of a marketing strategy. Therefore, professional advice regarding design and registration could prove to be a useful investment for the future.

9.9 DESIGN REGISTRATION

You can normally register the shape, design, or decorative features of a commercial product if the product is new, original, or previously unpublished—or, if it is already known, the characteristics can be registered if they have never before applied to the product you have in mind. Generally, clear representations of the design must be submitted along with a statement of why the design is novel.

There are generally two sorts of legal protection for designs: design right and registered designs.

9.9.1   Design Right

Like copyright, *design right* arises without the need to take any formal action and covers the internal and external configuration of a product or design. The design must be recorded in a design document or actually have been made. A method or principle of construction is not covered by design right—only the specific article is covered. Design right gives protection against others making, importing, or selling an identical or substantially similar article. Design right gives protection for the design for a period of up to 15 years but no more than 10 years protection from when it enters the market. Furthermore, in the last five years of the design right term, anyone can apply for a *license of right*. This means that you must grant a license to a prospective competitor on terms to be agreed. If you cannot agree, the terms will be settled by a legal third party.

9.9.2   Registered Design

Registered design covers specifically the external visual appearance of an article. Design registration applies only to features that appeal to the eye, not the way the product functions. To register a design, you would normally apply to the design registry at the patent office, sending a specimen or photograph of the design plus the appropriate fee. If the design qualifies, a certification of registration is issued, which gives the owner the sole right to manufacture, sell, or use in business articles of that design. This protection lasts for five years and can be renewed for up to four further periods of five years.

The designs of certain types of articles, such as where the shape is dictated by the function of the article (i.e., sculptures, medals, and printed matter such as book covers, calendars, and stamps), are generally excluded from design registration and come under copyright law.

There is no all-embracing international registration for designs, which means that in most cases separate applications have to be made in each country in which protection is sought. If an application to register a design is sought in a country that has signed the PCT International Convention, it is possible to claim "priority" on the earliest application made to protect the same design in PCT countries. This means that the new application cannot be invalidated if, during the time since the original filing, the design has been published or articles made to the design have been manufactured and sold or if, in the same period, a competitor has produced a similar product. A set time limit of six months dating from the earliest PCT country filing is allowed for claiming priority.

9.10 LICENSING AND FRANCHISING

If you have a revolutionary invention but don't have the finances, skill, or time to produce it, what can you do? There are basically three actions you can take once the idea is patented:

- *Outright sale:* You sell the rights and the title of your patent to an individual or company.
- *Sale and royalty:* You enter into an agreement whereby you assign the title and rights to produce the product to another party for a sum of money and, in addition, you get a royalty on each unit sold.
- *Licensing:* You keep the rights and title but sell a license for the product to someone else.

The first two options effectively give control to someone else. The third course, licensing, is quite interesting in that the rights and title have been retained by the inventor. A license does not affect ownership of the intellectual property. Instead, it gives permission to do something that would otherwise be an infringement. Licenses vary, but essentially there are four general types of licenses:

- *An exclusive license.* This means the licensee and the licensee alone has the right to do whatever the license allows. The owner is prevented not only from granting any other licenses, but from doing whatever it is the licensee has been licensed to do. The scope of the license can be restricted to a particular application, country, or market. It is therefore possible for there to be several exclusive licenses from a single intellectual property right.
- *A sole license.* This is similar to an exclusive license, but it does not restrict what the licenser can do. A sole licensee has no special rights in the event of infringement.
- *A nonexclusive license.* This permits the licensee to do things that would otherwise be an infringement on the owner's rights. Only the owner of the property is in a position to defend the right against infringement. For instance, it could be possible for Company X to be paying royalties on every component produced under license, while Company Y, which did not bother to take out a license, uses the same process for free. Unless there is something in the license agreement that obliges the licenser to take action against Company Y there is nothing Company X can do about it.
- *A sublicense.* This is a license granted by a licensee. License agreements may give the licensee permission to allow others to exercise the same rights without reference back to the original owner. Sublicensees are in an even weaker position than licensees when it comes to enforcing their rights because they are two or more stages removed from the owner.

Licenses can be on any terms to which the parties agree. They do not have to operate on a worldwide basis, but can be limited to certain defined areas. They can be for limited proposes, giving the licenser considerable scope to develop a strategy for exploiting the patent to the best effect.

Franchising is a certain kind of licensing. It has become a very effective business system and a fast-growing sector. It is mainly well regulated and supported by the financial institutions. Franchising can take place between the manufacturer and retailer, whereby the franchisee sells the product directly to the public. Examples of franchises include car dealerships, food chains, and gasoline stations. Franchising may also exist between manufacturer and wholesaler, whereby the franchisee produces a product under license (such as in the cola drinks industry), or between a wholesaler and a group of retailers with agreement on pricing and marketing.

A franchise involves one company or individual (the franchiser) granting a license to another company or individual (the franchisee) that allows the franchisee to trade under the franchiser's name and benefit from the goodwill and initiative of the franchiser. The main elements of a franchise are:

- A contract containing all of the terms agreed;
- The franchiser must initiate and train the franchisee in all aspects of the business;
- The franchisee must make a substantial capital investment;
- The franchisee must own his business;
- The franchisee is given a geographical territory in which to be a sole operator;
- The franchisee pays the franchiser for the rights that are required and for continuing services.

Taking out a franchise when starting a business reduces the risk, but profits are generally lower. From the franchiser's viewpoint, franchising offers the opportunity for expansion more rapidly with lower capital investment and less risk.

9.11 JOINT VENTURES

In addition to licensing intellectual property, it is possible to use the property in conjunction with other commercial relationships. The most frequently encountered is the joint venture. This is where two or more independent organizations with complementary skills and expertise join together to exploit their shared opportunity. Because legal implications over partnerships can be quite serious, contracts are always drawn up by professional experts.

A joint venture company is a company like any other, but what makes it different is the special purpose for which it is formed, which will be reflected in the way that its articles of association will be drawn up. The articles are the internal rules of the company, and they determine the nature of the relationship between the company's shareholders. The main features are:

- *Control.* Who is to be the majority or dominant shareholder? Can arrangements be made to ensure there is a blocking mechanism to control the dominant member?
- *Costs.* Can mutual pricing arrangements be made when materials or other inputs are being charged by one of the members?
- *Profits.* Can agreement be reached on how to divide profits?
- *Termination.* Will there be restrictions on the buying and selling of shares which hamper a member leaving the venture?

Pooling patents or other forms of intellectual property is a way of creating a collection of these rights to consolidate cooperation between different owners or interested parties. The contributors to the pool take any rights from it they need. Sometimes there is a need for a pooling agreement to proceed in a joint product or process. It can also be used by the parties as a way to reduce actual or potential competition in a trade or industry.

Package licensing occurs when related patents or rights are licensed in a block by the owner for a uniform royalty that must be paid irrespective of the actual use that the licensee makes of any individual items in the package. The justification for package licensing is, generally, that it avoids complicated accounting procedures for royalties and the valuation of the individual patents or rights. Pooled rights are often then licensed as a package to third parties.

Cross licensing of patents is a simple form of patent pool. For instance, in the development of the thermionic valve, inventor X held a patent for the two-electrode thermionic valve. Inventor Y discovered and patented an improvement to the patented invention of X. The invention made by Y was a big step in the development of the valve. Each inventor by virtue of their patent could have prevented the other or anyone from commercializing the new invention. Clearly, the patent of X placed him in a commanding position in the technology. He could continue to manufacture and market two-electrode valves, while Y could not have exploited his grid invention without X's license. An agreement between X and Y covering the exploitation of their respective inventions was in the best interests of all parties including the public, as long as the agreement was not likely to restrain trade.

Allowing partners to gain a monopoly or an unfair advantage over competitors is a difficulty that can arise from the pooling of patents. In the past, some industries did have a policy of pooling patents and rights, with the aim of attacking outside competitors by means of patent litigation. Therefore, parties considering cross licensing have to make sure that the venture and agreements do not violate anticompetition laws. If the patents pooled are complementary (as in the case of the thermionic valve), then a liberating rather than restricting effect should occur. Without pooling, some products would not be developed or would take a great deal longer to get to market. Patents, design rights, or confidential know-how can all be pooled when complementary and not lead to restraint of trade. For example, a good case for a pooling arrangement would be where inventor A has a patent for an alloy and a process for working the alloy, together with related know-how. Inventor B has a patent for a furnace that is ideally matched for working A's invention. Bringing A and B together would be very beneficial.

9.12 AGREEMENTS

Any parties interested in exploiting technology together will need to enter into formal agreement on the terms and conditions for exploitation. They will need to ensure that the needs of each party are acknowledged and accepted in a written legal document. The exact nature of the agreement will reflect the initial discussions between the parties, the specific requirements each has, and the rights/obligations each party is willing to uphold for the term of the agreement. These outline terms are commonly referred to as the *heads of agreement*. Before any of the parties begins the drafting a formal document, each party would first discuss what items it wishes to include under the heads of agreement and how it intends to deal with each item.

Once the parties have identified the particular form of agreement that suits their needs, they can then address the heads of agreement, which will usually fall under some or all of the following categories:

- Confidentiality;
- Duration;
- Well-defined work schedule;
- Reporting obligations;
- Liability;
- Payment schedule;
- Nature of the intellectual property (patent, copyright, know-how, etc.);
- Option term on license;
- Scope of license;
- Lump sum/royalties;
- Patent costs;
- Ownership of intellectual property;
- Ownership of improvements;
- Infringements (who takes action);
- Termination;
- Laws of the country.

Once the parties are satisfied that the heads of agreement accurately reflect the purpose of the agreement and their respective requirements, the heads of agreement are used as a template to draft the formal legal document. Most technology-transfer agreements include references to some form of licensing.

9.12.1 Consortium Agreements

This is essentially a joint-venture agreement between research partners who intend to jointly conduct a research and development project and exploit the intellectual property rights emerging from the project. The parties are usually a mixture of staff from universities, research organizations, and industry partners. The consortium agreement is used whenever two or more partners are involved in developing technology resulting from an independently funded research project. It is a separate document that is dependent on the research agreement between the partners, but which deals with the specifics of commercialization of the intellectual property. The draft agreement would be signed prior to commencing the research project and would provide for aspects such as:

- Defining the purpose and scope of the agreement;
- Appointment of a project management committee;
- The means with which intellectual property ownership or license rights are to be dealt;
- How patent applications are to be filed and associated;
- How costs are dispersed;
- Terms for maintaining the confidentiality of information and dealing with requests for publication.

9.12.2 Distribution Agreements

An agreement between a technology manufacturer and a distributor or agent in a specific country or territory is a form of technology transfer. This allows for the placing of a finished new product in a new market. A distribution agreement deals more specifically with sales and marketing conditions rather than the use and exploitation of technology as in license agreements.

In the agreement, the market territory (which can be a country or geographic region) will be defined. The agreement will also state whether the distributor is appointed on an exclusive or nonexclusive basis. The term of the agreement will be established. This is often structured in the form of a renewal of the contract for certain periods of duration. These are subject to a term of notice in the event of either party wishing to terminate the agreement. The price and terms of delivery are set. There is usually a requirement for the distributor to provide annual purchase forecasts and to agree on minimum purchase requirements.

The manufacturer would be expected to supply the distributor with sufficient quantities of the product (with a strict specification of quality or performance) within a specified time period. Technical backup and support for the sales team, as well as marketing, promotional, and support material and services may also be obligations of the manufacturer.

In the drafting of distribution agreements, the parties must take care not to include clauses that are anticompetitive in nature or an abuse of a dominant marketing position.

9.13 DISCLOSURE

In all aspects of intellectual property, a problem arises when an invention or new idea is first disclosed. For an invention, the first disclosure should be in the patent application. Many applications have been rejected because the invention has already been disclosed. Disclosure to the public domain can occur more readily than most people might consider. Publication in books and scientific journals are often the main focus of concern with regard to disclosure. Disclosure can, however, also occur in ways that the inventor might consider irrelevant. Short presentations are often made at seminars or workshops. Although posters or presentations at conferences, exhibitions, or local societies are similarly minor, they still constitute disclosure to the public domain. In particular, one must remember that the presentation or material is often reproduced in a booklet of proceedings of the meeting.

Material provided for publication in local newsletters is public communication, as is the placing of a thesis in a university library. If the thesis contains confidential information, then the author should place on it a formal restriction of release. A formal restriction of release allows the thesis to be made available to a third party only by approval of the supervisor. Therefore, the submitting author should make sure that such a system is in operation in the university library. The author also needs to be certain that any information that the library automatically enters into its interlibrary-loans database does not contain the confidential information.

Oral discussion and the passing of information between colleagues and potential commercial partners also constitutes disclosure to the public domain unless it has been made clear to the receiving party that the information is confidential (e.g., by the signing of a confidentiality or secrecy agreement between the parties).

A disclosure of an invention relies on the nature and degree of the information that has been communicated. The information varies in its relevance to disclosure, depending on how simple or how complicated the specific invention is. Generally, if the information provides a third party with an understanding of how the invention might work, where it could be applied, and how it might be repeated, it is a disclosure. Information ranges from a full scientific publication or the publication of significant results regarding a chemical structure or formula, a piece of electronic circuitry, an engineering concept, or a drawing, to the demonstration of a prototype device. A notice or advertisement to the effect that a material, method, or process is available for purchase can indicate that the information is available to the public domain. Thus, consultation with a patent agent or the person responsible for intellectual property in a company is advisable before any information relating to an invention is made available by any means to a third party.

9.14 AGENCIES

The individual or small company looking to exploit a piece of intellectual property or identify potential licensees or joint-venture partners faces what seems a daunting task. Equally, companies looking for new technologies to exploit face many problems in the complex world of searching the databases of publicly and privately owned research institutes and universities. Nearly all countries have witnessed a growth in the number of publicly and privately owned organizations whose business is to act as intermediaries in these situations. These agencies may be able to help with the identification of outlets for or sources of new technology or other intellectual property and help in a patent application. Their role is to act as a broker: they will not normally become directly involved by taking a license or any other interest in the property itself, but will draw on their resources, contacts, and expertise to give advice. They will do this for a fee—which is the way they earn their profits from the technology business.

9.15 CONCLUSION

The need to take intellectual property rights seriously cannot be over emphasized. The material presented in this chapter is a small extract of a complex area of law. A major point is to develop a rational strategy for the exploitation of your property, which should be part of the business plan. If you are planning to sell the patent to another party to exploit, the whole aspect of filing, timing, tactics, and licensing must be carefully planned. Today products rapidly become obsolete; thus products have decreasing lifespans. If the details of the product or product operation are published too early, competitors can render a product obsolete or have a near copy ready before or shortly after the launch of the original product.

References

[1]   Convention of the Grant of European Patents (European Patent Convention), Part II, Chapter I, Article 52.
[2]   Convention of the Grant of European Patents (European Patent Convention), Part II, Chapter I, Article 54.
[3]   Calvert, R. P., *Encyclopedia of Patent Practice and Invention Management*, New York: Reinhold, 1964.

Selected Bibliography

Bainbridge, D., *Intellectual Property*, London: Pitman, 1994.

Black, T., *Intellectual Property in Industry*, London: Butterworth, 1989.

Jacob, R., and D. Alexander, *A Guidebook to Intellectual Property, Patents, Trademarks, Copyright and Designs*, London: Sweet and Maxwell, 1993.

Jorde, T. M., and D. J. Teece (eds.), Antitrust, Innovation and Competitiveness, Oxford, England: Oxford University Press, 1992.

Matsushita, M., *International Trade and Competition Law in Japan*, Oxford, England: Oxford University Press, 1993.

Raybould, D. M., and A. Firth, *Law of Monopolies, Competition Law and Practice in the USA, EEC, Germany and the United Kingdom*, London: Graham and Trotman, 1991.

Rose, V. (ed.), *Common Market Law of Competition*, London: Sweet and Maxwell, 1993.

Chapter 10

Prototypes and Product Development

10.1 INTRODUCTION

So far we have considered the commercial and technical consequences of marketing a new product or service. Yet is there a product? Is it a generalized concept, still on the drawing board, partly constructed or a working model. Is it a system that can be simulated on a computer? If it is a service, can it be tested with customers? If it needs building, who makes it? A decision also has to be made on whether to use a prototype.

10.2 TO PROTOTYPE OR NOT TO PROTOTYPE

The company will need to assess whether to build a prototype. Not building a prototype generally allows a saving in time to actual production and a saving in startup costs, but building one can highlight product development problems. A lot of effort can be expended converting a prototype into production. Even if there is a prototype, the company must assess how long it will take and how much it will cost to get products into the marketplace. Generally, though, new-product development managers would state that, if in doubt, minimize the risk and uncertainty by spending the time to make a prototype, even if it is a nonworking example.

For one project, a Western manufacturing company recently bought some high-speed equipment from a Japanese manufacturer in addition to developing some machinery on its own. In the early stages of the project, while the Japanese firm was still investigating the products, the Western company kept reminding it of the schedule plan, which had a firm completion date. The Japanese company continually responded that it knew the schedule but was not ready to start production yet. The outcome was predictable: although development for the products produced by the Western company started earlier, some were over budget, late, and took longer to get into the working stage. The Japanese equipment was produced within the budget, on time, and took less time to arrive at the production phase. This is a lesson in patience and the concept of thinking about and working on the development phase of a product while establishing production procedures.

Decisions are made continuously as the prototypes move through various stages of development, but if there any any doubts about development, it is best to go to the next stage. For instance, in car manufacturing, after a sophisticated CAD image has been created, a car body in clay can be fashioned to see what it will look like "in the flesh." If the manufacturer is not content with this prototype, a "body in white" working prototype that closely resembles a a production-grade vehicle can be made.

One of the common problems associated with production is the use of nonstandard components during the building of the prototype. Only a small number of these nonstandard components is needed for the prototype, but will sufficient quantities be available from stock by the time production starts? For instance, if a specialized microchip has been used, will this chip be available to the desired specification and reliability, and in sufficient quantities, by the time production of the product commences? Also, will its cost be higher or lower?

Prototyping should be considered for use in product development because it can assist in the following areas:

- Planning of implementation;
- Checking on feasibility;
- Clarifying necessary requirements;
- Exploring various options;
- Improving understanding of the product;
- Improving understanding of development issues;
- Verifying designs;
- Exploring maintenance issues.

There are also some risks associated with prototyping, such as:

- Increased time spent on development;
- Insufficient design documentation;
- Making prototype too early;
- Insufficient testing;
- Lack of knowledge of the new technical environment.

The following checklist can be used to determine whether it is advantageous to prototype:

- Specifications and requirements are known overall but not necessarily in full detail;
- The benefits outweigh the risks;
- The new product is capable of being prototyped;
- The product to be developed is a substantial undertaking with large costs of development and implementation;
- The cost of the prototype in relation to the overall cost of the project is low.

A number of common problems occur in the process of prototyping:

- Developers spend too much time exploring nonessential variations of the prototype;
- Changes in the prototype have major effects on the implementation;
- Changes in the prototype drastically increase the overall costs of implementation and/or the cost of the product on the market;
- Problems in scaling up from a prototype to a production model are not fully assessed.

There are no definite rules for the decision whether or not to build a prototype. Only guidelines can be offered. Experience can be an essential ingredient in deciding which way to go, with respect to the knowledge of the development of both the concepts and the production processes.

10.3 PRODUCT DEVELOPMENT

A major factor in the success of the adoption of new technology is the product development. It is well known that it is getting more difficult for a new or changed product to be successful in today's marketplace. Customer requirements, tougher legislation, and competitive forces mean that products developed by traditional design- or engineering-led methods run too high a risk of missing their targets.

These traditional methods are often described as "over the wall" in that marketing departments would identify an opportunity, pass it over the wall to the design department to map out a suitable product, and then manufacturing would produce it. This is all done with the minimum of consultation and the maximum likelihood of expensive late reworks of the design. Reducing product development is an increasingly important order.

The success of the Japanese, in particular, in reducing the time taken to bring new products to market has led to an increased interest in techniques such as concurrent (or simultaneous or parallel) engineering.

Table 10.1 shows a typical comparison of the times for stages in product development for a Japanese and a Western company.

Table 10.1
Response Time in the New Product Development of Mechanical Transmissions

Process	Time Taken to Complete (Months)	
	Japanese Company	*Western Company*
Develop design concept	2	3
Complete layout and design	1	3
Design review	1.5	1
Detail design	1	4
Manufacture prototypes	2.5	6
Pilot test and power test	1	2
Field test	4	8
Manufacture first product	4	7
Total time	*17*	*34*

Table 10.1 illustrates the significant difference in the total time from concept to first product. In fact, with this comparison the Western company takes twice as long—in a competitive world, this is disastrous.

10.4 CONCURRENT ENGINEERING

The world markets are demanding products that are well designed, of high quality, and at low prices, with ever decreasing lead times. Thus, the pressure is on manufacturers to utilize the best methods of technology with optimum control and efficient management, along with suitably enabling organizational structure. Concurrent engineering (CE) is becoming accepted as the methodology that can help satisfy these demands and maintain the profitability and viability of product developers, manufacturers, and suppliers. CE has been defined as the simultaneous development of a product and the process required to produce it. A fuller definition provided in [1] is:

CE is a systematic approach to the integrated, concurrent design of products and their related processes, including manufacture and support. This approach is intended to cause the developers, from the outset, to consider all elements of the product life cycle through to disposal, including quality, cost, schedule and user requirements.

CE involves participation of downstream functions, including manufacture, supply, quality assurance, and customer support. The objective is to produce an optimized design incurring minimum cost over the life cycle of the product. Successful implementation of a CE strategy depends more on the culture of communication and participation within an organization than on the degree of automation in the production process. That information exchange is seen as a key factor in CE suggests that companies will need more open organizational structures, such as teamwork and matrix management. CE is therefore the integration of all of an organization's assets needed for product development, including people, tools and resources, and information. The main benefits of CE are:

- Decreased product development time;
- Greater competitiveness;
- Close integration between departments;
- Improved product quality;
- Greater control of design and manufacturing costs;
- Promotion of team spirit;
- Improved profitability;
- Enhanced reputation of the company and its products.

A successful demonstration of the benefits of CE can be seen at Lloyds Instruments of Fareham, U.K. They brought their latest materials-testing machine to the market in June 1993 after just seven months of preparation. This represented a halving of the time for the previous machine coming to market and was a major competitive benefit for a company with sales of £5 million and a workforce of 75. The company installed a network of computers in its factory that allows instant access to the latest design drawings for all departments in the company.

Other examples include Rolls Royce, which reduced the lead time to develop a new aircraft engine by 30%; ITT, which reduced its design cycle for an electronic counter measuring system by 33%; and McDonnell Douglas, which reduced production costs by 40%.

The 1992 Design Council/EDS-Scicon survey of 11,000 U.K. readers of the journal *Engineering* painted a depressing picture of the actual uptake of CE and an even more depressing picture of products late to the market.

The lessons of such recent surveys point to the need to adopt the marketing-oriented policies discussed in Chapter 2. The starting point for new product design should come from a close relationship with customers and a knowledge of their needs.

For example, the U.K. company Tensator moved from an engineering-led to a customer-led approach in the late 1980s. It produces pedestrian guidance systems consisting of posts and retractable tapes commonly used to form queues in post offices and banks.

Its philosophy now features:

- Regular feedback meetings with regular customers;
- A computerized database of customer information and economic data;
- Computerized CE access to design drawings;
- Product development meetings of staff (and some suppliers and customers) in *cross-functional* teams;
- Control of the stages;
- Checks that each stage is completed;
- Decisions regarding which other management functions may be needed to be brought in and when;
- Estimates of required resources.

Lloyd Instruments is too small to have cross-functional teams—the same people would be in every team. Instead it holds regular product development meetings led by the engineering department and involving the managing director, senior design staff, and representatives of sales and production. The meetings are briefings of progress, and actions are identified. For instance, production can make tooling decisions while marketing can work on advertising material.

Unlike Tensator, Lloyd Instruments does not invite its suppliers or customers to its meetings, but it has found extra benefits by contracting out more design work to its suppliers.

The greeting card industry does not necessarily come across as ripe for a revolution in new-product development. With birthdays and holidays coming around every year, it is surely a question of having products that fit in with general taste.

This is not so, according to Hallmark Cards. To continue to grow at the rate that had characterized its 80-year history, the company needed to expand its market. Unfortunately, new-product development was taking three years from initial concept to store shelf. The solution for the company was concurrent engineering/cross-functional teams, and development time was slashed to less than a year. This was done as part of an overall *reengineering* of the company—another management buzz phrase (which will be discussed more in Chapter 12), but one that produced major quality/productivity gains for Hallmark.

10.4.1 Concurrent Engineering Management

Terms like *cross-functional team* are borrowed from the language of total quality management (TQM). CE as a whole is very similar to the TQM concept of quality function deployment (QFD).

With QFD, a team drawn from all of the functional parts of an organization works together to bring the advantages of:

- Promoting teamwork;
- Improved communication;
- Better structure for information;
- Better frameworks for analysis;
- Living documentation available to everyone.

Therefore, using CE or QFD does not eliminate the need for a subtle structure of planning/control/management. A typical set of the management stages is as follows:

1. *Create a design and development planning program.* The complexity of the requirements will determine the need for and the extent of this program. Plans are typically presented in chart form showing the different activities against a timescale.
2. *Provide design practice codes and procedures.* The structure and control of the system for managing design should be fully described and documented. This should also include control of specifications.
3. *Assign the design activities.* The planning of design should include the provision of resources and qualified, trained staff suitably housed and equipped to carry out the assigned tasks.
4. *Identify organizational and technical interfaces.* The work of designers must be controlled when it affects the work of others, including subcontractor design activities.
5. *Identify the design input requirements.* The inputs required for a new design of products or services would be clearly defined in terms of performance, use, aesthetics, technical details, packaging, storage, reliability, maintainability, and disposal.
6. *Identify the design output requirements.* The design process should produce the technical information for purchasing, production/operations, and inspection maintenance in a clear and comprehensive manner. This usually comes in the form of a specification, which would not prescribe irrational tolerance levels.
7. *Establish design review procedures.* The objectives of design reviews are to ensure that the design meets the requirements specified and that progress is being made towards the objectives. This will require a system for the timely identification of problem areas.
8. *Investigate new techniques.* Monitor the balance between innovation and standardization. *Evaluate new materials under the appropriate conditions.* Investigations and evaluations are necessary to establish reliability and safety implications of untried methods and materials.
9. *Use feedback data from previous designs.* Procedures would be established to ensure that the valuable experience gained from previous designs and customer experience was put to use in future designs. Thus, from here one could go back to stage 1.

10.4.2 Design Review

It is worth emphasizing the *design review* CE management stage. This is used to ensure that the process of design progresses toward the objectives of the program. The aims of the design reviews are to establish that

- The design of the product or service will meet all the specified performance criteria;
- The product or service being designed can be produced, inspected, tested/checked, installed, operated, and maintained to provide satisfaction to the customer;
- All possible design alternatives have been considered;
- All statutory requirements will be met;
- There is adequate documentation to define the design and how the product or service is to be used and maintained.

Design reviews should cover all of the quality-related factors, and checklists of these should be drawn up for each review. These are usually carried out at the preliminary, intermediate, and final stages of the design process by either representatives from all of the relevant disciplines or by a QFD team.

10.4.3 Rapid Prototyping Processes

We have discussed various aspects of the management of CE, but what are the most important tools to help CE? It is generally accepted that these are the use of 3-dimensional (3D) computer-aided design (CAD) and the various rapid prototyping techniques. Designers have been able to use CAD to fully define the shapes of objects. The use of computer numerically controlled (CNC) machines allows the production of objects with great accuracy. However, there are many items with complex geometries and small features, which make machining very difficult or impossible. These problems can now be overcome by using what are known as rapid prototyping techniques. For greater details of such techniques we recommend [1].

Basic rapid prototyping uses the geometry created during surface or solid modeling of the CAD to produce the prototype model. The designs are transformed into a physical 3D model by adding layers of material, one on top of another, until the complete part is constructed. To be able to transfer the CAD data to the rapid prototyping machine, the model is tesselated so that the surfaces are covered with a mass of triangles. The sizes of the triangles depend on the curvature of the surface and the resolution required. The finer the resolution used, the smoother will be the surface produced. This, though, will be at the expense of huge increases in the amount of data needing to be transferred. The rapid prototyping processing techniques in use at the moment are as follows:

- *Stereolithography.* This is based on the use of a light-sensitive polymer fluid and sophisticated laser scanning.
- *Selective laser sintering.* This system is like stereolithography but uses a powder in place of the polymer fluid.
- *Fused deposition modeling.* The machine uses molten plastic similar to nylon or casting wax.
- *Solider.* This is a combination of stereolithography, fused deposition, and a CNC milling machine.
- *Laminated object manufacturing.* This is a different approach from the others, as the object is built up with sheets of material such as paper or plastic. The sheets are stacked on top of each other and adhere to one another. After each new sheet is placed in position, a laser cuts the outline of the part.

There are many examples of companies such as Porsche and 3M using rapid prototyping to reduce lead times and produce significant cost savings. In the new environment of CE, larger effort is put into the early design stages, but this extra cost of time and labor is being more than paid for by the savings in later activities.

Finally, it is worth noting that rapid prototyping does not just occur in manufacturing. There is now what is termed the rapid prototyping of software and hybrid systems. This does not quite fit in with this section on CE, but for details of this activity we refer you to [1,2].

10.4.4 Key Features of Concurrent Engineering

The U.S. Institute for Defense Analysis investigated the deployment of CE in organizations. The key features that they identified as essential elements for successful implementation of CE are:

- Multidisciplinary teams;
- Computer simulations of products and processes;
- Sustained communication and coordination across different disciplines and companies involved with the product;
- Integration of databases, application tools, and user interfaces;
- A commitment of continual improvement;
- Use of quality-management methods and principles;
- A program of education for employees at all levels;
- Employees developing an attitude of ownership towards processes in which they are involved.

So the accomplishment of CE requires integration, coordination, and management of computer-based tools; creation of an organizational structure for team working; evaluation; control; and the appropriate use of new technologies. Also important for the best results are the considerations of internal and external customers and suppliers, as well as the synchronization of data and information.

10.5 REVERSE ENGINEERING

We have discussed the state-of-the-art methodology for product development with CE. Yet if a company is to implement CE, it will usually need to find a considerable investment from somewhere. For some companies this is not possible. The economic recessions have forced them to lower rather than raise their levels of capital investment in equipment and staff. Thus, the production systems need to be maintained in their present condition for longer periods of time. As a result, the decline in a working production system will continue unless preventive or corrective action is taken. These companies, however, still need to progress and compete in the marketplace. One of the solutions to this problem of maintaining a system for a longer period of time is reverse engineering (RE).

RE is basically the development of the technical data necessary for the support of an existing production item developed in retrospect as applied to hardware systems. An excellent text on this process is [3]. RE is a four-stage process:

1. Evaluation and verification;
2. Technical data generation;
3. Design verification;
4. Project implementation.

RE begins with the components that have been prescreened against specific criteria. The components are singular parts, such as a circuit card, or the smallest complete unit of a system, such as an electronic module. The RE process identifies and strengthens the weak links in any system. The process can be depicted as:

product → disassembly → measure & test →
design recovery → prototype & test → RE product

whereas the more traditional process is:

need → design idea → prototype & test → product

The benefits claimed for RE are an increased ability to maintain a manufacturing capability at peak performance rate. From this, enhancement and improvement of the end product will stem. These then produce savings through the use of improved product documentation. The savings may be invested to upgrade the system. So, where modernization of a system is too costly, RE can act as a stop-gap function to increase system productivity until profit can be made available to invest in new technology. RE is a tool used to improve productivity, but it is not just for manufacturing. It can be used to improve any product line, service, or capability within an organization. It effectively helps those who have older systems compete with those who have newer technology by sharing and developing technical information.

10.6 SERVICE INDUSTRIES

It might seem strange to think of prototyping in the service industries. Yet companies involved in this sector have to respond to change and be innovative just as much as those in any other commercial sphere. Because starting a new service often requires a large investment, prototyping a service can be more critical than prototyping a product. The benefits, advantages, and risks in prototyping for a service are very similar to those for a manufactured product.

If you are interested in prototyping in the service world, you may wish to research the idea of *proof of concept* used in the U.S. This is where a company forms a service that works like its final service will, but which uses completely different underlying technology to make it work quickly and cheaply. This is then tested by potential users and customers, and feedback and support for the project are obtained. As an approach for further prototyping, the company then uses the technology that it intends to apply in the final service. This technique is often used in the computer industry—particularly in systems development. Further details of the practical use of proof of concept can be found in [4–6].

A major objective of the companies in industries such as food and drink, airlines, and retail is being competitive and having an advantage over competitors. This results in these companies continuously looking at product development. In the service sector, product development and its methods are very diverse. For instance, the managing director of one brewery developed and modified the company's range of beers by spending a great deal of time talking and drinking with customers. British Airways has a policy of constant development of its services, as shown in Section 10.7. One important means of achieving this is through the use of customer and staff feedback. McDonald's introduces new food concepts by first testing them alongside present brands in selected takeout restaurants in certain geographic areas. The leading producer of turkey foodstuffs in the U.K. has achieved market dominance by being innovative in the development of new products such as turkey burgers and turkey "dinosaur" drumsticks.

Companies can develop new products (to them) by taking over other companies to diversify or enhance their own product range. For instance, a newspaper publisher whose paper has successful sales Monday through Saturday may take over an ailing publisher of a Sunday newspaper to attack that market. The introduction of new technology can be used to improve and speed up the service to customers. The U.K. retail giant Argos, which sells a wide diversity of household goods mainly in large superstores, claims that 96% of products requested will be in stock. It believes that customers will not tolerate goods being out of stock. Therefore, the company feels that it is necessary to have the most up-to-date computer system possible to meet this need for customer satisfaction in its service. Argos had about 50 million customer transactions in 1992. To meet this demand, the computer system has to work quickly and effectively and ensure efficient stock processing through electronic data interchange links with suppliers.

There is wide scope for product development in the service industries, but the number one objective is staying ahead of the competitors and getting a bigger market share.

10.7 EXAMPLES

10.7.1   Example A: Company X

Engineers at Company X designed a revolutionary compressor for its refrigerator products, and a futuristic factory was built to make it. A new powdered metal was used to produce the compressor, which eventually contributed to the failure of this compressor. The two powdered metal parts were excessively burning up oil and eventually choking off all cold air to the refrigerator. What happened to cause the company to go into production without knowing about the problems?

Apparently, in the eagerness to reduce costs, field tests had been reduced, other testing had been accelerated, and the compressor was rushed into production. Poor communications contributed to the disaster. Several of the employees who did the preproduction testing suspected that the compressor might be defective and did in fact tell their superiors. Unfortunately, senior executives did not hear of the employees' concerns or the potential problems that existed. The new compressor seemed ideal; it weighed less, was more energy efficient, had fewer parts, and was to have a manufacturing cost about half that of the older reciprocating compressors. There was enormous pressure to bring the new compressor in on schedule to begin the cost savings earlier. In 1981, a decision was made to drop the new compressor and use purchased ones, five of them from foreign producers. In 1989, Company X revealed a $450 million pretax charge resulting from the unsuccessful compressor development program. Good lessons exist here on not going too far too soon.

Not enough analysis was done at the development stage, and communications were poor with management not listening to its workforce. The company did not see who could provide the materials as well as themselves. In fact, there were failures along the whole decision process, which resulted in a very costly disaster and is an example of very poor technology transfer.

10.7.2 Example B: The City Bee

The City Bee is a prototype electrically powered, zero-emissions car. It is being developed for production by a consortium of companies from Norway, Austria, and Switzerland. The prototype was tested as a post office vehicle at the 1994 Winter Olympic Games in Lillehammer, Norway. The original idea came from the Norwegian company Bakelittfabrikken A/S which manufactures the car's thermoplastic body in one step using a new and innovative technique of rotation molding. The surface does not require painting and is extremely tough. A blow with a heavy hammer on the surface supposedly leaves no mark. The car has been designed to use an aluminium frame and fewer components than conventional cars. Thus, the car is very light (weighing about 400 kg), is especially quiet, and has a top speed of 100 km per hour. It can carry two people and some luggage for a journey of 50–70 km on one battery charge. This compares well with present cars in the countries developing the new car. Tests are continuing with the car, but it is hoped that if the field trials are successful the car will be launched on the market in 1996.

10.7.3 Example C: The Integrated Product Development Centre

In most countries, small companies often lack the new-product development skills required for successful growth. In Scotland, many small companies lack the practical knowledge to innovate. Concurrent engineering was being used by the multinationals—most of the recorded practice and CE tools related to large companies. So the Design Council in Scotland created a new initiative called the Integrated Product Development (IPD) center. This was aimed at small companies, particularly in the Strathclyde area, to bring together the expertise of local universities and large firms to produce practical guidelines for smaller companies. The center was based on the CE research center at West Virginia University in the U.S. The university discovered that if companies do not treat product life cycle issues early in the design process, problems are discovered too late, and long and costly design iterations are required.

The IPD center intends to provide a focused, formal approach to new-product development techniques. The first stage is to initiate pilot projects to investigate how CE tools more usually associated with large companies could be adapted to small manufacturing companies. On completion of the projects, a comprehensive package will be presented to interested companies. The work will then be documented to encourage local small companies to use concurrent engineering methods and technologies. Many of the companies targeted are suppliers that are increasingly expected to perform product-development and added-value roles for their clients—something that Scottish companies have traditionally found difficult.

10.7.4   Example D: Almhults EL Mek

We introduced the case of Almhults EL Mek in Section 4.2.2 of Chapter 4. The company wanted to develop a lifting magnet.

The lifting magnet was developed in cooperation with the University of Vaxjo. The University helped the company with market research and developed a specification based on customer requirements.

Toward the end of the development process, the company tested a prototype with three different companies. These companies analyzed and measured usage. These results were then used to make further improvements. By this process (i.e., where the future customers' experience and knowledge were used), several advantages were ascertained:

- The product was developed to the needs of the customers.
- During the time of development, potential customers were contacted.
- Experience gained from the testing with customers could be used in marketing the product.

The elm lifting magnet became the first product fully developed by Almhults EL Mek. This has since been followed by other systems including magnet separators for nonmagnetic materials.

10.7.5   Example E: British Airways

British Airways' policy is to strive for total customer satisfaction. To achieve this objective, the company believes in a continuous development of its service product. Customer satisfaction on certain key indicators is monitored on a monthly basis. Ongoing research attempts to identify customer segments and their expectations. Nearly 200,000 interviews are carried out each year with passengers and travel groups. Some of the improvements made stemming from this research are speedier check-in facilities, express check-in for shuttle services on domestic routes, and onboard portable computers for cabin staff to make the buying of duty-free goods quicker and easier. Improvements have also been made to the food and drinks served on flights. The company believes that to continually improve the quality of service, the management must listen to comments by staff. So they offer financial incentives, as much as thousands of English pounds, to staff for suggestions on how to improve efficiency. The company has also formed staff groups whose role is to propose ways to improve services to customers.

A major aspect of assessing the service is the company's system of quality measurement and communication. Employees work with some 200 separate performance targets, both quantitative and qualitative, that are produced for the managers who then proceed in a way they think is best.

So product development need not be merely producing something new every few years—it can be continuous development of a particular product, as with this airline. This is often the case with the service industries.

10.7.6 Example F: Thorntons

Thorntons was established about 80 years ago with one toffee shop in Sheffield, England. Now Thorntons is the largest chain of independent shops supplying toffees and chocolates in the U.K. The company is particularly well known for its brand of freshly made luxury chocolates. Thorntons has always strived to produce confectionery that is different from that of its competitors and of a high quality. It has about 400 shops in the U.K., of which about 200 are franchises (usually in towns too small to merit a full Thorntons outlet). Such franchises are in shops or supermarkets where selling space is limited to generally one counter.

In the confectionery market, the company has been quite innovative in its product development. In-store displays were used during the industry's off-season to encourage sampling of new products, such as wine-flavored truffles. In 1988 the company purchased the Belgian confectioner Gartner Pralines, which specialized in high-quality chocolates and fresh cream products. Pralines sold its products mainly through patisseries. Thorntons is now distributing Praline's products through its own U.K. stores and selling Thorntons products through the Belgian company's outlets. In 1989 Thorntons purchased Sogeco and La Nouvelle de Confiserie, which have particularly well-known brand names in Europe. In the constant push for new products, the company is looking at producing a sugar-free range and the "super choc," which would be one range to suit all Europeans (something difficult to achieve on any front).

10.8 CONCLUSION

There is no set of rules regarding the question of whether to build a prototype. In effect, every company innovation is by definition unique. The emphasis in this chapter has been on the need to consider the technical implications of a new product or service. It is essential to consider these implications as part of the overall business plan. Any new product or service is bound to have an element of risk. The aim is to minimize the risk as much as possible. Prototyping can and generally is a major element in technological evaluation and assessment of risk. Is there more, though, that can be done to reduce the risk of failure? The answer is, of course, yes and we will discuss this in Chapter 11.

References

[1]   Syan, S. C., *Concurrent Engineering: Concepts, Implementation and Practice*, London: Chapman and Hall, 1994.

[2]   Rahmani et al., "Rapid Prototyping via Automatic Sotware Code Generation from Formal Specification," in *Behavioural Models Specifying User Expectations*, J. A. Kowal (ed.), London: Prentice-Hall, 1992.

[3]   Ingle, K. A., *Reverse Engineering* London: McGraw-Hill, 1994.

[4]   Birkmayer, W. S., and M. J. Wale, "Proof of concept model of a coherent optical beam-forming network," *IEE Proceedings Part J Optoelectronics*, August 1992.

[5]   Shonder, J. A., "Proof of concept of a magnetically coupled Stirling engine-driven heat pump," *Proc. 27th Intersociety Energy Conversion Engineering Conference*, (PIECDE 0146-955X).

[6]   Ting et al., "100MeV proof of principle laser Warefield accelerator experiment," *Proc. of IEE Particle Accelerator Conference*, Vol. 4, 1993.

Chapter 11

Reducing the Risk of Failure When Introducing New Technologies

11.1 INTRODUCTION

Any technology transfer or innovation project is associated with uncertainty. There may be technical risks (failure to chose the correct idea, inability to keep within resource/time targets), commercial risks (failure to meet market needs), economic risks (the financial market may change during the lifetime of the project), timing risks (too early a launch, too late a launch), and human risks (poor management, lack of staff competence/motivation).

Radical innovations may bring great uncertainties about their outcomes but the benefits will often be correspondingly great. Also, as Chapter 12 will explain, there may be an even greater risk in not innovating.

Almost any of the planning or good management practices described earlier in this book will reduce the risk of failure for the project. General good practices that reduce the risk of failure would thus include:

- Maximizing market research/technical intelligence monitoring;
- Having the closest possible customer and supplier interfaces;
- Having financial appraisal techniques that are suited to technology transfer–based initiatives;
- Routinely assessing the communication routes in the organization;
- Sufficiently training staff in team building, team facilitating, empowerment, and new technical skills;
- Routinely checking the project criteria and progress against targets and against evolving company strategies;
- Having proper project termination criteria in place;

- Developing a balanced portfolio of innovative programs that cut across different types and varying levels of risk;
- Drawing on internal strengths and the past experience of successful transfers of technology;
- Increasing the number of new projects to increase the chances of a major success.

This chapter aims to describe certain techniques that identify or assess specific areas of uncertainty or risk.

11.2 SCREENING

The concept of screening criteria for new technology projects is a useful overall approach that embraces a lot of decision-making steps and audits described elsewhere in this book. The screens can be used either as an aid or as a weapon. As an aid they can help in ensuring that successful projects are chosen and completed. As a weapon they can be used by senior management to ensure that low innovation or personal agenda routes are chosen. Clear guidelines, based on genuine organizational belief in the process, can maximize the chances of fair decision making.

General screens should be based on the company's mission and strategies. For example, early screens for a technology project in an upmarket company might be: does the project maintain the quality image of the company's products? Is it environmentally friendly?

Financial/business analysis screens usually follow the general screens. For example, the NPV targets in Chapter 7 were typical financial screens. Unrealistic returns on investment (ROIs) or very short payback periods show the problems of aggressive screening criteria.

Marketing screens might be:

- Difference from competitors' products;
- Uniqueness;
- Quality level;
- Whether the changed product can be delivered through existing channels.

Technical screens would be the sort of evaluation criteria covered in Chapter 6.

Well-designed screening criteria therefore provide a consistent method (with limited senior staff bias) to compare suggested technological changes, either with other possible changes or with the status quo.

11.3 RISK ASSESSMENT

Many commercial software packages that allow quantification of the relative risks in decision making are available. Certainly the detailed approach in these packages is recommended if staffing and other resources are available, but even if time is limited it is possible to carry out worthwhile risk assessment.

This belief that some simple analysis is much better than doing none will be stressed again in the next two sections of the chapter.

The first step of assessment could be the identification of all sources of risk in the new-technology project. This procedure will, of course, in itself be valuable in avoiding surprises, but a mental perspective will also be gained from thinking about interrelationships in the project. A team approach to the assessment exercise has the benefit of revealing different perspectives on company's strategic objectives or project priorities. However, an interesting psychological phenomenon called the *risky shift phenomenon* must be taken into account. It has been observed that a group of people is more likely to go with a riskier decision than each member would individually—the fear of responsibility is hidden within the collective responsibility.

Sources of project character risks might include:

- Level of management experience;
- Whether the technology is proven;
- Whether the technology is accepted in the company;
- Correctness of cash inflow projections;
- Correctness of cash outflow data;
- Effect of project on overall company survival;
- Effect of failure on other projects;
- Likelihood of meeting the time schedule;
- Ease of monitoring the project progress;
- Commitment level of management.

Sources of general business risk might include:

- Market potential;
- Market price change;
- Competition;
- Financial scenario changes;
- Environmental risks;
- Regulation risks;
- Internal personnel risks;
- Information accuracy.

The second step of assessment is to ensure that the lists of risk are subjectively assessed (say on a scale of 1 to 5) in terms of their risk level. The total gives a useful indicative risk figure.

Also, the risk scores for each source of risk can be divided into either controllable or uncontrollable, and a ratio of controllable to uncontrollable can be calculated. It may be that some uncontrollable sources might be outsourced to other organizations or perhaps shared with suppliers.

A detailed description of this simple risk-analysis method is given in the excellent article by Groth [1]. We have used Groth's method with success in a number of project situations.

11.4 SCENARIO PLANNING

In any business, closing one's eyes to the need to make decisions is in essence a form of passive decision making. It is an acceptance that the business will be allowed to take its chances on a "keep going as is" basis. If nothing in the environment changes, this may work perfectly well, but the assumption of stability around the business is questionable for most enterprises.

Strategic planning should look at longer term issues whose impact is likely to be far reaching or enduring. Those who do not plan may simply crash into an immovable barrier that could have been circumvented or accommodated. If the desirability of strategic planning is accepted, the question of handling uncertainty in that planning must be considered. In the 1990s there is a lot of uncertainty inherent in the future, and dealing with it is a hard task for management.

There are many approaches to uncertainty, and several are related to the idea of probability. However, as the planning horizon lengthens, the assessment of probabilities is increasingly difficult. Also the assessment of long-term probabilities is very subjective. To deal with these complexities, scenario planning can be used. The sort of scenario planning used by the multinationals is, of course, too expensive for smaller companies, but some basic techniques can be successfully used by anyone.

The references at the end of the chapter contains details of another excellent article, written by Foster [2], that we have used with success in practical situations. It contains examples of some general economic scenarios. The description that follows here draws on Foster's work.

According to Foster, a scenario can be defined as a description of a possible future, based on a set of mutually consistent elements, within a framework of specified assumptions.

The basic steps of writing a scenario are shown next:

1. *Key factors.* Identify key relevant factors—economic, social, technological, political, and competition—inside the organization and in the external environment.
2. *Assumptions.* State the assumptions to be implicit in the scenario, including its horizon.
3. *Sources.* Identify relevant sources of information for key factors.
4. *Issues.* Identify the issues/points of divergence arising from conflicting forces in the current situation.
5. *Pictures of the future.* Assemble or fit together the information and forecasts obtained to generate internally consistent pictures or development pathways.

Step 5 is easy to write down but will be potentially difficult in practice to handle. The assembly process can lead wholly or in part to a computer model but, for a small organization, could be a largely descriptive process annotated with a few key items of hard data along the way. One way of examining the potential interactions of factors (and future forecasts of them) is to resort to *cross-impact analysis*. At a smaller company, this may amount to constructing a simple matrix with the factors on both axes, thereby focusing attention on combinations that staff may then evaluate subjectively.

Complementary issues that arise in practice are:

- At what level of generality/aggregation should the scenario(s) be pitched?
- How many scenarios should be built?

The two most commonly used levels of scenario are global scenarios and industry scenarios.

Global scenarios usually deal with global economic issues. Typical major variables are: growth, inflation and interest rates, protectionism, and energy prices. Usually it is large organizations that use this type of scenario (e.g., those in the aerospace industry).

Industry scenarios typically include factors relating to competitor behavior and new technologies.

For the smaller business, global scenarios may not be relevant and industry scenarios may be at too high a level of aggregation. Instead *local market* scenarios can be produced, with issues similar to those of the industry scenario but more narrowly and locally based.

On the question of how many scenarios to attempt, the theoretical answer is "a lot," but in practice this is not feasible, even for large organizations. If practical considerations mean that the scope is limited to, say, two scenarios, it has to be decided which two scenarios to attempt. Four possibilities are shown here:

1. *Most likely.* In this scenario, trends unfold according to our best estimates. Some factors may still change sharply, because events do not evolve in a continuous way.
2. *Surprise free.* In this scenario existing values and trends are maintained so that we face no "uncharted waters."
3. *Optimistic.* This is the ideal scenario, where everything is good.
4. *Pessimistic.* In this scenario, everything goes against the company, subject to the usual constraint that all of the elements must be internally consistent. Financial variables, competition, and other factors go against the company.

If only two can be attempted, it might be advisable to choose 1 and 4. The first scenario might be selected because it is where we expect to be, and the fourth might be chosen because it is the one we have to explore to find out if we can stay in business if the worst happens. The ideal scenario is probably least useful—if that set of events occurs, the worst that can happen is a failure to exploit the opportunities available. This may be irritating, but it is hardly life threatening. If it is possible to do more than two scenarios, the surprise free model may be helpful because it provides a *base scenario* from which to move outwards—it might be thus useful despite its low likelihood of occurrence.

Large-scale planners might dispute the use of one of a small set of archetypal, thematic developments because putting all of the good things into one scenario and all of the bad things into another might not be internally consistent. But in the real world of a hard-pressed or under-resourced organization, the smaller scenarios can still be of great benefit.

Once the scenarios are built, the company's position can be analyzed and decisions can be made on a course of action. The choice is between designing the strategy around the needs of a given scenario or trying to hedge the strategy (i.e., devising a strategy that will be workable across a range of scenarios). Hedging may be unnecessarily cautious. Probably the best option is to devise a strategy to optimize within the most likely future but with contingent escape routes built in.

11.5 FAILURE MODES AND EFFECTS ANALYSES

A good technique for checking both the product design and the proposed manufacturing process is that of failure modes and effects analyses (FMEA). FMEAs are a useful tool for objectively analyzing the design, where possible failure mechanisms, their causes, effects, and prevention are brainstormed by the project team. Two analyses are carried out, one on the product design and one on the process. In each, reference is made to the product specification and the relative importance of the various characteristics. Ranking of the severity (*S*), likelihood of occurrence (*O*), and probability (*P*) of detection take place for each set of failure modes, effects, and causes and a *risk priority number* (RPN) of the product is obtained. The highest ranking can be addressed in further development.

The following is an example of the instructions written for the project teams for FMEA in an electronic ceramic manufacturing company.

11.5.1 Failure Modes and Effects Analyses for XYZ Company

FMEA is a controlled system for predicting what may go wrong with a device or process. It takes into account the design, manufacture, and working life of parts and processes, and tries to identify every potential failure. The assumption is that failure occurs, however unlikely, and the objective is to establish how it occurs and what the customer would find objectionable (excluding failure to meet specifications).

Possible failures are analyzed according to:

- The effect on the customer;
- The severity of the effect;
- The cause of the failure;
- The frequency with which it is likely to occur;
- The probability that the failure will be detected.

It is then possible to establish a program of corrective action.

If it is not possible to completely eliminate the cause of the failure, and therefore the failure itself, the FMEA can be used to indicate what can be taken to reduce the possibility of the failure occurring and increase the chances of spotting the fault before it reaches the customer.

The FMEA is to be carried out as a team exercise with input from design engineering, manufacturing engineering, production, quality, and other relevant personnel.

Thus, an FMEA is a tool enabling its user to systematically:

- Identify potential failures;
- Evaluate them;
- Investigate them;
- Take action to prevent them.

The design FEMA. The purpose of the design FMEA is to eliminate the design aspects that will result in failure of the component or product. The assumption must be made that all of the components have been assembled correctly and received to specification.

Failure modes. The question to be considered is "How could this part fail?" The team must consider every way possible, however unlikely, in which the product, component, or assembly can fail to perform its intended function, anticipating failure of unspecified functions as well as those for which there exist defined tests. It is necessary, at this stage, to find ways in which it could fail, not whether it will fail or not (e.g., the terminals of an automotive sensor might fail salt spray test requirements).

Effects of failure. For each failure mode there will be at least one effect. These are to be stated and described in terms of what the customer might notice or experience (e.g., the effect of disc material drift over life to the extent that failure occurs is that the device being switched does so at an incorrect temperature).

Causes of failure. Each failure effect has at least one cause, and it is the purpose of this section to identify them all, regardless of their likelihood. The more complete that the list of failure causes is, the easier it is to pinpoint the problems and consequently to eliminate the failures (e.g., the design causes of a loose terminal may include incorrect dimensioning or incompatible rivets).

Rating failure. On completion of the list of failure modes, effects, and causes, the factors are rated to obtain the risk priority number (RPN). The RPN is the product of the failure mode's severity, occurrence, and detection, as described here:

1. *Severity.* The severity of the failure is a measure of the importance of not failing. It is classified on a scale of 1 to 10, where 1 is a trivial nuisance and 10 is life threatening. Severity of any one failure will be allocated over a range depending on the application, but it must also take into account the importance of the failure to our company.
2. *Occurrence.* The question "how likely, in reality, is it to fail?" is considered. The response is classified on a scale 1 to 10 to determine severity.
3. *Detection.* The team estimates the probability, using existing controls, of detecting the failure. The 1 to 10 scale is used again; for a rating of 1, a near 100% probability exists that the defect would be found, and for a rating of 10, the defect would not be seen until the customer finds it.

(Note: Generally, there would be individual company guidelines accompanying these notes to help with the allocation of the scale).

Risk priority number. The RPN is the product of severity, occurrence, and detection scales multiplied together. The higher the RPN, the more serious the failure, effect, and cause. A critical summary is obtained to highlight areas where most action is required. The use of the RPN, therefore, eliminates those failures that are unlikely to occur when evidence of their existence was demonstrated. In other words, design priorities are established. Failure modes where severity is 9 or 10, however, must always be addressed, regardless of RPN.

Recommended action. After establishing a reasonable list of failure causes associated with the highest RPNs, action must be recommended to eliminate them.

Each action must be summarized, with ownership allocated and a completion date given. Precise statements of action, rather than vague, meaningless suggestions, ensure that everything is clear and that actions are carried out on time.

Follow up. The FMEA or project leader is responsible for making sure all recommended actions are completed and documented by the target date. However, it is the actioning person's responsibility to carry out the work and to document the result.

After completion of the corrective action, a new FMEA must be carried out, taking the changes into consideration.

Controls and checks used in conjunction with an FMEA must be incorporated into manufacturing procedures and specifications for the product (e.g., if the FMEA assumes an acceptable quality level, this must continue into the process FMEA, control plan, manufacturing, and QA instructions).

On completion, all issues of the FMEA must be available, together with all documentation regarding recommended action. The FMEA is a live document and must be reissued upon any change in design.

(Note: There would be a very similar document for a process FMEA incorporating all the sections as per the design FMEA.)

11.5.2 Notes on FMEAs

All FMEAs are subjective and ratings allocated cannot be compared with those obtained from FMEAs carried out externally to the company. FMEAs contain information regarding manufacturing techniques and procedures that are confidential to the company. It is prohibited, therefore, to release process FMEAs to customers or any third party. It is permitted that noncompany personnel may view process FMEAs only in the presence of a member of the FMEA team, preferably the engineer in charge, and only when an approved meeting, either at the customer's premises or at the company site, is in process. Design FMEAs tend to be carried out for custom-designed products and therefore may be released to customers for their information.

The XYZ Company example shows that as a procedure, FMEAs are a excellent control mechanism even though they may appear laborious. FMEAs also help with the next stage of investigating new techniques and materials, where the exercise, based on current practice and the peculiarities of the new techniques and materials, can be analyzed. Unfortunately, FMEAs are difficult to establish because of the mental disciplines required. However, they are useful and become easier to carry out as team members become more practiced at them. Software packages that make use easier are available, but not all are user friendly. These software packages are advertised in engineering journals, and it is best for the users to determine the one most suitable for their applications.

11.6 CONCLUSION

Innovation and technology transfer cannot exist without uncertainties and risks. Although time for considering the sources and scale of these uncertainties and risks is generally seen as a luxury, techniques like risk assessment, scenario planning, and FMEA have major benefits for a company.

First, they give quantitative (if sometimes subjective) measures that allow for relative ranking of alternative technologies/projects/courses of action. Decisions have to be taken, so it is best to make these decisions as safe as possible.

Second, and equally important, the process of carrying out the techniques forces staff to find out all of the sources of risk, to think about the issues involved, and to work together, revealing possible differences of understanding of the project or company strategy.

References

[1]   Groth, J. C., "Common-sense risk assessment," *Management Decision*, Vol. 30, No. 5, 1992, pp. 10–16.

[2]   Foster, M. J., "Scenario planning for small businesses," *Long Range Planning*, Vol. 26, No. 1, 1993, pp. 123–129.

Chapter 12

The Management of Innovation and Change

12.1 INTRODUCTION

Hopefully any company that appreciates the major recommendations of the previous 11 chapters of this book will also embrace the importance of innovation in general and technology transfer where required. But a step beyond embracing the importance of innovation is becoming an organization that genuinely has an innovation culture (i.e., one that goes beyond putting innovation/technology transfer in its mission statement but feeling unable/unwilling to do much about it). The 3M Corporation had a five-year objective in 1990 to generate 25% of its sales from products developed during those five years. The target will be met as a direct result of the corporate structure (cross-functional teams, communication routes, etc.) and beliefs (e.g., the staff is allowed to pursue research projects of their own choosing).

However, as examples in Section 12.4 will show, having world-class innovation levels (Philips in the late 1980s) or well-planned use of new technology and teamworking (Chrysler since 1990) are not sufficient if nontechnology planning fails. Philips could not properly link innovation to customer requirements in the market. In 1995, Chrysler faced a hostile takeover because it had chosen to maintain a high cash reserve (against a possible future market downturn) instead of passing on benefits to shareholders. It is so hard to get everything right!

12.2 COMMISSIONING A NEW PRODUCT

There is a plethora of books on "managing change," "strategies of technological management," and "the management of innovation," and all have their own style and advice on how to succeed. Of course, each company's requirements will be different, and each needs to define its own strategic approach to management to achieve its goals. It is perhaps useful, though, to highlight some of the key factors common to most theories and to look at some examples.

First, let us look at a comparison between what are thought to be the traditional and the innovative managerial emphases. These comparisons between the traditional approach, with a more closed culture, and the more dynamic open organizations are being stressed in many of the articles and books on managing innovation. The reason is that, by a large factor, the majority of the successful companies who have introduced change put the newer, more open approach to management as a major parameter for success. In diagrammatic representation, the approaches look very similar, but note where the emphasis is stressed within the two cultures in Table 12.1. Also consider the communication flow; in the open system it is by far more a two-way process. It has now been realized that people are a very important resource in the strive for success and profit. Hence the introduction and push in the U.K. of the "Investors in People" award for companies. This requires a company to show in its strategy the role and importance of its personnel in the production or service process. There are two other main differences between traditional and innovative companies: the need to look continually at new options and opportunities and the need to put customer requirements as a major or even top priority.

Table 12.1
Comparison between Traditional and Innovative Companies

Criterion	Traditional	Innovative
Culture	Closed	Open
	Directive	Explores new approaches
	Based on attitudes derived from past successes	Facilitating
Strategy	Rigid	Proactive
	Formal representation and justification of status quo	Focus on opportunities
		Identification of desirable changes
Systems	Inflexible	Flexible
	Optimized to run a tight but inflexible ship	Able to accommodate change but still maintain a tight ship

Criterion	Traditional	Innovative
People	Servants	Initiators
	Perform tasks allotted to them	Business involvement
Orientation	Internal	External
	(avoiding internal disruption)	(meeting customer needs)

After: [1].

In the traditional organizations, communication is generally limited and almost entirely *top down*, whereas in the innovative organizations, the volume is high in both top-down and *bottom-up* directions. A company based on a belief in technology, growth, and the need for change is more likely to come under the heading of *innovative*; in fact, most of the features of the innovative organization are found in successful technological companies, particularly in Japan.

To encourage innovation, the most important of the factors that an organizational structure needs to take into account are:

- Achievement of the long-term development of staff and technological expertise against the short-term needs to apply the technology for commercial gains;
- Transfer of technological information into the company from outside sources and the transfer, if necessary, of corporate policies into R&D;
- The association of marketing, production, and financial staff with R&D staff, to an optimum extent;
- Preservation of a high degree of autonomy for project managers while maintaining corporate control over resources invested in a project;
- A leadership style appropriate to social and organizational needs;
- Recognition of the company's technological profile;
- The need to stimulate creativity and motivate staff.

This very much ties in with the findings of a study of high- and low-performing British companies over the past 20 years [2], which revealed five key factors needed for effective change management:

- Leadership by the management team;
- Constant evaluation of the changing commercial environment;
- Treatment of people as assets rather than costs;
- Conversion of strategic change into operational actions;
- Constant all-embracing coherence (i.e., a sustained sense of purpose).

Management seems to have little choice: being in business means dealing with change. Sir John Harvey-Jones (former chairman of ICI) is quoted as saying [3] "management is about maintaining the highest rate of change that the organization and the people within it can stand."

It seems that there will be no forward progress and no tangible improvements without change. However, this is not a new idea, and it is worth noting a quote in 1513 by Machiavelli [4]: "and one should bear in mind that there is nothing more difficult to execute, nor more dubious of success, nor more dangerous to administer, than to introduce a new order of things: for he who introduces it has all those who profit from the old order as enemies and he has only lukewarm allies in all those who might profit from the new."

12.3 THE RELATIONSHIP BETWEEN TQM AND CUSTOMER SERVICE

To obtain successful innovation management, the management of the company as a whole must be examined. It is possible to manage innovation separately from the other functions of the organization if it is narrowly defined as making "technical" changes. However, with a wider definition that encompasses new products, process improvements, marketing for increased differentiation from competitors, and organizational change for cost reduction and quality improvements, it is likely that the organization will use the same management practices for innovation as for all functions, treating it as an interlinked part of the whole management process. This wider definition is unashamedly taken from the total quality management (TQM) approach to business and has undoubted value throughout the organization. Taking a TQM approach to the business and its operations has a number of results.

12.3.1 TQM Results

Overall Quality Consciousness Throughout the Organization

What is quality? Is it giving the best at any cost or is it giving the customer what he or she really needs? If a customer wants a self-powered device to carry up to four people and baggage from one location to another, equipped with a wheel at each corner and a steering mechanism that allows control over the direction of the movement, do you supply a Rolls Royce or horse and cart? Both match this specification and thus are "quality" solutions, but the horse and cart are all that the customer really needs. The refinement and style of the Rolls Royce are not, unless these items are additionally specified.

With consciousness of quality (i.e., matching the customer's needs) the following are obtained:

- A clear business mission;
- Clear commitment of senior management to improve quality;
- Involvement of all staff in working together for the right product;
- An understanding of the importance of planning and communication;
- An understanding of the merits of problem prevention over their detection;
- Companywide commitment to constant improvement, change, and innovation.

Knowing and Reacting to Customer Needs

If the manufacturer of an item knows exactly what the customer wants to use it for, the company, as the expert, can improve and develop the product to further meet the needs of the customer. Communication thus plays an important role, for it may be that the customer has simply specified the nearest product offered, which will "do" until a better one comes along. Obviously, of the two options above, most people would prefer the Rolls Royce to the cart for matters of comfort, but, because of cost, they choose the cart. If a competitor introduces the Ford Escort at one tenth the cost of the Rolls but twice the cost of the horse and cart, it does not take a genius to work out whose business will decline!

This knowledge of what the customer really wants and the willingness to respond accordingly results in:

- Faster, more flexible, and improved product development;
- Better scanning of the business environment and better market research;
- Closer relations with both customers and suppliers.

Ensuring That Products Do Meet the Customers' Needs

Care and thought about the customers' applications ensure that they remain pleased and, therefore, will continue to do business with you. However, this also requires:

- Improvement of the quality of material from suppliers;
- Monitoring and control of products;
- Training.

For these purposes new methods of operation are often required (e.g., zero-defect philosophy and statistical process control).

Making the Customer Happy

In addition to meeting expressed needs (e.g., product specification), attention must be given to the whole package, including:

- Packaging, instructions, and presentation of product;
- On-time deliveries;
- After-sales service.

Paying attention to these details results in a delighted customer.

Overall Competitiveness

A single approach to business, followed by each section of the organization, is very effective. For example:

- Public relations and advertising should come from an overall marketing philosophy.
- Lower costs could resulting from production savings (e.g., smaller inventories, reduced scrap, better investment analysis).
- Performance measures and benchmarking should indicate areas for action.

The organization that recognizes these major points will probably succeed in its innovation as in its other functions.

12.3.2 Example: Simon Access

An example of the need for not only management to embrace TQM, but for all personnel to be involved and become committed to the customer-led approach, is Simon Access in the West Midlands of England. The company had secured itself a major-player role in the market for specialized access equipment, such as that used by fire departments, emergency services, and the construction industry. To continue operating at this level, particularly against tough Japanese competition, the company felt the need to instill a genuine customer-led ethos. Central to this was to use the innovative ideas of shop-floor workers and technology transfer, tied to tangible relationships with customers at all levels. An early, simple approach to informing staff was to take everyone on the payroll to a major exhibition of firefighting equipment.

The effect was dramatic in modifying long-ingrained attitudes, such as a perceived lack of competitors and an ignorance of the technologies used elsewhere. Also, for the majority of the workers, this was the first time they had ever seen a customer!

Going to the exhibition provided an impetus to the launch of a program called "Best of Both Worlds," which was designed to change the company culture and emphasize two company mission targets: to be seen as a major supplier by customers and to maintain a high degree of autonomy at local factory levels to increase flexibility and innovation.

Although long and expensive, the program was seen as essential for competitiveness, moving from a product-driven organization to one that provides solutions to customer problems. Training of senior staff to become fully committed to the new culture was intensive. Also, user groups were established with customers to give opportunities for constructive discussion, leading to product changes, information on new technologies, and a number of new product ideas.

With a new customer services manager brought in from British Airways and located at the production plant, targets and achievements are widely displayed and visiting customers are always brought to the factory floor to talk with workers. This stems from the world class performance (WCP) benchmarking approach used within the company, whereby the company has the aim of a 90% rating after an assessment against industry-best practices of 78 key business procedures.

12.3.3   Consequences of Insufficient Planning

To see what happens when aspects of planning or management are not fully considered, let us look at the collated results of several studies of European engineering companies with which the authors are familiar. After the introduction of new technologies by these companies, it was found that:

- At the end of the project periods, most companies reported that they had failed to meet their financial criteria, either because of unrealistic financial-appraisal techniques or because of organizational problems during the implementation.
- Almost all of the companies encountered bottlenecks due to a failure to create all of their organizational or information structures in time.
- Almost all had difficulty with supplier quality or delivery, often due to inadequate liaison.
- Half of the companies had difficulties with changes in the marketplace that they had not foreseen
- In most of the companies the process/manufacturing planning went well, reflecting a technologist-led approach. Preventative maintenance and operator training were correspondingly found to be inadequate for the new system requirements.
- Even when information-flow structures were created, they were often overly formal, inflexible, or limited in scope.

In summary, what these companies had failed to do was ensure that sufficient project planning had been carried out by someone capable of taking a holistic approach to the situation. In addition, the projects were started before the support actions were available. It also appears that there was insufficient market research, and communication, internally and externally, was poor. The TQM approach, on the other hand, should prevent these disastrous product introduction schemes from taking place.

12.4 MANAGEMENT CHANGE PROGRAMS

In larger companies, it is often the case that the need to move to an innovative or TQM culture is achieved as part of a complete business *reengineering* program for the organization. In 1995 Ford announced Ford 2000—"the Mother of All Reengineering Programs," designed to ensure their full global competitiveness. Major examples such as Chrysler (twice) and Goodyear have become very well known, but they are important. Chapter 11 reminded us that mistakes can still occur with the best reengineering programs, but these mistakes become problems rather than death blows.

The second Chrysler revival from 1992 clearly showed the basic tenets of reengineering and a number of these are technology/innovation related:

- Cross-functional, or *platform*, teams designed to break down old separate functional *chimneys*, such as design engineering, procurement/supply, manufacturing, and sales. There became just four platform teams instead: small car, large car, jeep/truck, and minivan.
- Faster and cheaper product development. Concurrent engineering reduced the total cycle time for the new Neon model to two years with development costs concentrated in a tightly directed period once the platform team had reached a consensus.
- Integration of most research staff into the platform teams (e.g., the electric vehicle research program was placed within the minivan team to ensure that new electric vehicles were market-driven as well as technology-driven). Only 69 central R&D staff were left.
- "Tech clubs" on common-interest subjects across the platform teams to improve horizontal information flow (and prevent duplication).

Training was another important factor in the Chrysler program and will be looked at in Chapter 14.

Goodyear, like Chrysler, came to its corporate change program from a position of desperation, after unwise diversification and huge debts. There were similar basic tenets in the Goodyear program:

- Good communication and clear objectives.
- Six main cross-functional units of "associates" (rather than "staff"). Empowerment of staff to have responsibility at relatively young ages (in some companies, it is possible for a staff member to buy a $250,000 house in their personal life but be unable to authorize expenditure of $5 at work).

- Relating technology transfer and innovation to customer requirements. Ten percent of time is allowed for self-directed research, but the remaining 90% is based on clear market-oriented functions.
- A definite commitment to faster new-product development through concurrent engineering (30% of sales from products developed in the last five years).
- Close monitoring of performance.

The case of Philips came from an different direction. Although the company was an acknowledged leader in consumer electronics, it was brought to a 1990 annual loss of $2.5 billion and 7,000 job cuts (out of 32,000). Too often its products were not what the market wanted or were too late or too expensive. Also, technology transfer was poor between divisions and between the company and the outside world.

One of the objectives of a new externally recruited chief executive in 1991 was to foster a business attitude among technology/research staff. Two-thirds of research funding would come from product groups rather than headquarters. Technology transfer was to be improved by forcing divisions to coordinate their five-year product plans with the laboratories and factories. Concurrent engineering and improved communications were to be prioritized. A technology and new business assessment committee was set up to allow anyone with a promising idea to bypass divisional hierarchies. Ideas that pass the screening stage could be pitched in front of senior company staff (thus also improving business-planning skills and entrepreneurship in general).

12.5 CONCLUSION

As stated in Section 12.1, it hard to get everything right throughout the company's activities. Your business strategy should be both thorough and incisive; you need a good information system (including knowledge of transferable technology); you need to understand the financial market; the customer must be delighted; you need to be good at motivating staff; the marketing must be just right; and that is only the beginning! To have an innovation culture seems just one more difficult skill to master. But a genuine innovation culture is as essential for long-term survival as any other skill required for modern business success. A quick basic list of rules for moving towards such a culture might be:

- Turn specialists into generalists and ensure that knowledge is pooled.
- Break down traditional hierarchies to put emphasis on experimentation and creation in place of a deadening emphasis on control.
- Ensure that there is a genuine belief at all levels of the company that some mistakes are inevitable in an innovative climate.
- Improve communication through the company.
- Allow time for thinking.

References

[1]   Twiss, B., and M. Goodridge, *Managing Technology for Competitive Advantage*, London: Pitman, 1989.

[2]   Lorenz, C., "Management study stresses teamwork," *Financial Times*, November 12, 1991.

[3]   Harvey-Jones, J., *Making It Happen*, London: Fontana, 1988.

[4]   Machiavelli, N., *The Prince*, Chicago: Chicago University Press, 1985.

Selected Bibliography

Johansson, H. J., *Business Process Reengineering*, New York: John Wiley & Sons, 1993.

Oakland, J. S., *Total Quality Management*, 2nd ed., Oxford, England: Butterworth-Heinemann, 1993.

Chapter 13

Project Management and the Launch of the New Product

13.1 INTRODUCTION

In the book so far we have examined the concept of technology transfer and the need for innovation. Chapters 3 through 8 outlined the external and internal analysis that a company needs to fully ascertain all of the information required prior to developing a new product or service.

For the remainder of the book, with regard to the technology-transfer task list, we are looking at the following:

- Establishing practical management systems;
- Obtaining technology-awareness training;
- Organizing product-specific training;
- Launch of the product, process, or service;
- Monitor and control of the total operation.

In this chapter we assume that the company is going ahead with the new or changed product, and we look at aspects of implementation within the organization.

In a recent British CBI/DTI study [1], the key ingredients for successful innovation were highlighted under the following headings and subheadings:

- Company culture;
- Internal processes:
  - Making it happen;
  - Progress and review;
  - Training;
- External influences.

These were considered the three groupings in which to discuss change and its success. These very much tie in with the technology-transfer process as presented in the chapters on the implementation phase. The remaining chapters of the book will embrace these in their content, but will use the terminology of the technology-transfer process as outlined in the first paragraph of this chapter. All of the successful practitioners of innovation and change accept that the success of implementation is dependent on good management systems, staff training, and the monitoring and control of the performance of the operation.

13.2 PROBLEMS OF IMPLEMENTATION

Lessons on securing a strategy that results in success can be learned from other companies, often from those that have failed in their business objectives. In 1988, the London Business School looked at the reasons for small business failure in the U.K. [2] and found the facts detailed in Table 13.1:

Table 13.1
Reasons for Small Business Failure

Total Sample	Primary Factor (%)	Contributory Factor (%)	Ranking Factor (%)
Undercapitalization	54	77	1
Poor operations management	29	57	2
Short-term liquidity	26	60	3
Poor chief executive officer	23	37	5
High gearing	23	50	6
Poor state of the local economy	16	35	7
Theft & dishonesty	11	18	9
Bad debts	11	27	10
Increased competition	8	40	11
Other personal reasons	8	9	12
Loss of vital personnel	16	17	13
Obsolete product	6	15	14
Other management reasons	6	8	15
Poor-quality product	5	21	16
Poor facilities & marketing	4	25	17

The key factors, which in many ways are intuitive (but obviously not to the companies that failed), were ascribed as lack of finance, poor management, lack of financial information, and poor marketing.

The ensuing research showed that failure to tackle these problems may not lead to immediate business failure but will identify an inefficient organization characterized by:

- The inability to meet objectives;
- Management spending too much time on "firefighting" problems;
- Poor company morale;
- Unprofitable or minimal returns.

The typical response of someone whose business has failed runs along the lines of "we ran out of money," "there really wasn't a market for our product," "our debtors wouldn't pay up," "the competition was too good," "the business was not well situated," "we were undercapitalized and staff costs were just too high." All of these negative problems would have been avoided with good planning, as already indicated in other chapters. In general, business failures are found to be caused by poor management and insufficient planning.

13.2.1 Example A: British Satellite Broadcasting

British Satellite Broadcasting (BSB) was intended to be Britain's first satellite television system after it was granted a license in December 1986. It had backing from such blue chip companies as Reed International, Granada Television, and Pearson (the owner of the Financial Times). BSB had, as part of its mission, "to outspend and outgun its rivals." The only rival was Rupert Murdoch's Sky Television, but unfortunately BSB made several fatal mistakes.

First, it threw away its advantage of being the market initiator. Each satellite system had its own dedicated receiver, which meant that the first company to sign the customers to its specialized equipment would hold onto subscribers. Even though BSB had an 18-month lead in license and financing, amazingly, Sky was operational 15 months before BSB. By April 1990, Sky had installed 750,000 dishes and BSB never caught up.

Second, it was late going on the air because it had concentrated on unique but untried technology, which did not work and had to be redesigned. Third, after losing its lead, it allowed Sky to define the market. Finally, its costs were simply too high. When the plan and its execution are the problem, big money cannot always solve it.

BSB effectively wasted its starting advantage by not exploiting a new market ahead of its rival. It failed to implement its strategy in the time required and, in March 1992 "merged" with (in reality, it was taken over by) Sky.

13.2.2   Example B: Wang Laboratories

Wang Laboratories was founded in 1951 by Dr. An Wang, a Chinese immigrant to America. The company focused on mini computers and custom systems, but Wang failed to understand the changes brought about by personal computers and software developments. It was too slow to respond to the demands of the marketplace and, even though it had sales of more than $3 billion in 1985, just seven years later it was seeking bankruptcy protection, after shedding nearly 60% of its staff and facing debts of around $400 million.

Wang had fallen victim to a rapid change in technology and markets in an industry where change is the norm. Management had failed to respond to market changes and the challenges presented by competition. Unfortunately, the company needed to develop a new product but appeared to spend more of its time, energy, and resources defending an outdated market position.

13.3 SUCCESS STRATEGIES

In an ideal company, strategic plans for the business as a whole and for the divisions or operating units within it will already include the role of innovation—the exact detail and its relative importance depending on the company or unit itself. In the real world, however, these do not exist and the management team needs to determine its policy on innovation in the context of its business strategy, current technology trends, and from competitors' behavior. As shown in Chapters 6 through 10, management will already have made many of the crucial decisions in the technology-transfer process, summarized as "Yes, we're going ahead with product X," and then broken down into questions such as "Is the company going to make or buy critical components? Which technologies will be sourced inhouse? Will we embark on licensing, joint ventures, or acquisition? Will we use universities or other sources of intellectual property?"

It is necessary to formulate the strategy for successfully implementing the project, determine the steps necessary to maintain and strengthen inhouse skills, and ensure that progress of the overall plans, particularly in assessing the commercial gains, is monitored. Implementation of an innovation plan can be broken down into the areas of strategic management, training, product launch, and performance monitoring.

A strategic planning system is typically composed of several subsystems, which include:

1. A system of plans or the output of the planning process (what we are going to do);
2. The planning process, which defines the approach taken by management in developing the system of plans (how we got to step 1);

3. The decision subsystem or approach to be taken in finalizing decisions regarding the plans (what we used in step 2);
4. The management information subsystem, consisting of relevant data with which decisions are made (the basis of the information we used for step 3);
5. The facilitating organizational structure that describes how the organizational roles are defined (the people we used in step 4);
6. A planning management subsystem that describes the strategy and approach for implementation of the planning process (how we made the plans in step 1).

The implementation of these plans results in sound management systems. Again, this is very company-specific but we would suspect that, for most technological companies, these management systems would include:

- Strategic technology management;
- Project management;
- Production teams;
- Simultaneous engineering teams;
- Other project teams;
- Manufacturing technology;
- Marketing technology;
- Financial management;
- Management information.

Any change in the organization or its mission, objectives, goals, or strategies will affect the existing culture. For example, managers introducing the development of new products by cross-functional design teams will invariably find that the operating characteristics of these teams will begin to influence the culture of the organization and the following might happen:

- Greater participation in decision making by those previously not expected to do so;
- Improved communication channels;
- Improved product and process development with better quality;
- Reduced hierarchical expectations.

Although this is a theoretical example, it demonstrates that the ability to manage an organization's culture is very important. It is very much a Japanese priority to be successful in such management, and as we see on inspection of the electronic and automotive markets in particular, this has produced high commercial returns for that country.

13.3.1 Examples

Nissan Motor Company

For many years Nissan produced poor profits mainly because the company produced poor-quality cars, but it is now marketing an excellent range of vehicles. This turnaround came about by a radical change in management style and company culture, led by the most senior managers. Some of the new ways of doing things that affected the workforce are:

- Eliminating the seniority-driven culture in which younger employees were reluctant to speak out on issues until their seniors had voiced their opinions;
- Allowing employees to own competitors' cars and encouraging them to do their own competitive analysis;
- Introducing flex-time work schedules and standard dress;
- Opening a company house for single employees at the suburban Tokyo Technical Center.

Employees became involved in studying competitors' cars, participated in design changes, and noted trends that might affect new car design. The major lesson learned was that a company's culture cannot be radically altered unless senior management first admits that there are problems and then publicly leads the way.

Honda

Every year, Honda's R&D personnel spend up to two months in the field with customers. This experience gives them a sense of the market and customer preferences, together with knowledge of competitors' products and strategies. As a result, this reinforces the corporation's customer-oriented culture.

Tektronix

In the early 1980s Tektronix was a leader in the manufacturing of electrical and electronic test and measurement equipment. The company was also the world's leading maker of oscilloscopes. It was well positioned for the future as a leading supplier of color graphic computer terminals. However, it began to miss out on computer-aided design workstations and personal computers, and in some cases the company was being beaten by some of its own former employees. Tektronix had been successful and was considered the best place to work—in fact, key characteristics of the culture included participatory management, an egalitarian atmosphere, generous day-care center facilities, and flexible time off. These factors should have contributed to a favorable culture for success, so what caused the downfall? In addition to these good factors there were some overriding bad elements of culture: an old-boy network of top executives and financial "bleeding" due to inadequate cost controls. In addition, in general, it was "too comfortable a place to work" (this does not imply that there should be a "sweatshop" or slavery approach, just that there was no desire or real need to be productive). All of these factors led to competitive disaster.

13.3.2 Problems in Implementation

It is said that implementation is the process to convert strategies into realities. Certainly it is a stern test of the management team's ability to lead and manage. Its main components are:

(a) Leadership;
(b) Communication;
(c) Action plans;
(d) Budgets.

Before discussing the successful way forward, it is worth noting and pondering on the ten most frequently encountered problems in implementation. In order of importance they are:

1. Implementation took more time than originally allocated.
2. Major problems that were not identified beforehand surfaced.
3. Coordination of activities was not effective.
4. Competing activities and crises distracted attention.
5. Employee capabilities were insufficient.
6. There were inadequate training and instruction for employees.
7. There were uncontrollable external factors.
8. There was inadequate leadership and direction by the department managers.
9. Key tasks and activities were not defined in enough detail.
10. The information systems for monitoring were inadequate.

All of these points are addressed, or should be, in the components (a) through (d). The key to it all, though, is successful project management.

13.4 PROJECT PLANNING

The successful implementation of a new-product introduction relies on its planning and the management of the resulting project.

- Leadership styles vary but effective leaders create a climate for progress and increase a company's energy level. A leader shapes and shares a vision that gives a common point to the work of others. Leadership is not just that of an individual but the collective influence of the management team.
- The need for communication is self-evident, in that it mobilizes the workforce behind any development plan, creates enthusiasm, and maintains morale, although practical ways of obtaining effective communication are often hard to find. Publicity of ideas helps people to see the need and logic for change, and knowing about the future prevents worry about specific changes because it aids thought on how individuals can contribute to, rather than fight, the plan.
- The action plan, or project plan, is the means by which the strategic development plan is implemented.
- Budgeting and planning are essential partners when carrying out technology introduction because budgetary control is crucial, particularly for smaller companies (see Table 13.1).

At this stage it has been decided that the company will go ahead with expansion by means of a new product and how the company will develop the technology. By far the easiest way is to utilize inhouse expertise by creating a project team made up of members of every department that will be affected by the change. Minimum requirements would tend to include members from R&D, product design, production engineering, production, marketing, and quality, although the exact makeup of the team, obviously, depends on the nature of the company and its business.

One of the most debated topics in innovation management is that of self-managed teams, and major examples of team-based reengineering of organizations (such as Chrysler and Goodyear) were discussed in Chapter 12.

The key principle is one of empowerment, where employees are fully responsible for their work. The impetus for teams came from the need to make companies more responsive to the market and more cost effective, as well as the need to reduce cross-functional divisions. Within the team, the role of manager moves from director to mentor or disappears altogether.

Employees are expected to contribute ideas and recommendations; sometimes their jobs are redefined and, as a team, they decide how to do the work. Supervisors merely offer support. The team decides how it will work and solve problems, and the employees together run the unit as a business.

If the teams work correctly, they run themselves and need no boss. The team sets schedules and targets, and it orders materials and equipment. Its size can be anything between 3 and 30 members, with a mix of "worker" and "management" functions. Often the team is given responsibility to hire and fire members, looking after and improving quality and nurturing its customers.

In practice, although the general idea is for the team to operate as a self-managing unit, it will always need to be overseen. Company management must allocate one team member, often the most senior, as the project manager. It is the project manager's responsibility to ensure that adequate planning has taken place and that the team, in agreement with the plan, is working according to it.

13.5 GOOD PRACTICES OF PROJECT MANAGEMENT

A well-planned project is all that is required to introduce a new product successfully and easily into the company. The trick is in the planning and management.

A checklist for basic project management might be:

1. Establish clear requirements of the functions of the project and the team members. Check that all team members are capable of carrying out their appointed tasks. If necessary, carry out a training program.
2. Prepare outline plans and budgets (e.g., "We intend to launch the product at the trade fair in October next year, and we need a budget of $500,000 to cover development and manpower costs").
3. Carry out feasibility study; prepare a prototype at this stage, if one has been decided upon.
4. Prepare a formal proposal for approval by senior management. It is this information that will be used by management to secure the project's finance.
5. Obtain approval and establish companywide priorities for project tasks.
6. Allocate task types to each team member—these are usually obvious (e.g., all tasks related to marketing are carried out by that department's representative). But it is particularly important when a department (e.g., purchasing) is not represented on the team. To successfully complete all tasks, it is important to establish who does what and who liaises with whom.
7. Carry out a detailed project plan and budget and agree on start and completion criteria (e.g., "The project is completed at the first full-scale production run and will be formally handed over to production at that time"). Beware of being too optimistic—prepare an "ideal world" plan, where everything is achieved the first time it is attempted, then add contingencies to allow for every eventuality. Brainstorm possible failures, even the most ridiculous ones, like Martians landing and stealing the components. Eliminate them through risk analysis, supplier control, and common sense. As a team, draw a project network from which the project plan in the form of a bar chart of action versus time is obtained. This is not only a method to monitor progress, it also externally demonstrates intent to the team and can be used to obtain support from colleagues. There are number of good software packages on the market that

ease the project planner's burden and aid progress monitoring and reporting as the project proceeds.

8. Communicate the plan and budget, obtaining departmental approval and agreement of priorities for the use of the representative for the project, where necessary. The budget is of key importance for approval by financial management to ensure that sufficient funds are available: remember those key reasons for business failure! Communicate the intended start date; this is often forgotten in the enthusiasm to get going.

9. Monitor and control progress. Establish a cycle of reporting, task allocation, and progress meetings. If there is any slippage of the project's tasks, inform management of the consequences and investigate recovery plans. If good planning has taken place initially, there will be sufficient time available from contingency or by *critical path compression* (i.e., shortening the project time by adding resources in key areas or by overlapping tasks).

10. Document the progress and ensure senior management remains fully aware.

11. Document the formal handover to the management in charge of daily operation of the new product.

The following are common reasons for project failure, which can be avoided through good planning and common-sense management:

- Communication:
  - Targets were unclear.
  - Targets were not broken down into what had to be done and when.
  - There was no feedback on how targets were achieved.
- Motivation:
  - Progress and problems were not shared with the team, and no help was expected or given.
  - Too much time was spent on documentation without clarifying what was being recorded.
  - Too much time was spent on analysis of blame for missed targets rather than recognition of targets met.
  - Project members did not see themselves as a team.
  - Top management did not appear committed to the project.
- Organization:
  - Project manager's authority was not defined.
  - Project team members' jobs were not defined.
  - Project tasks were not clearly allocated to people.
  - It was not clear if the project manager or department head supervises the tasks.
  - There was no clear organization to arbitrate on decisions of priorities.

A key element in carrying out any project is communication: if in doubt, communicate it. Ensure that progress meetings are just that (i.e., team members are reporting their progress, not lack of progress). Any problems must always be communicated immediately to the project manager for contingency plans to be formed or corrective action carried out. Do not wait for team members to inform you of a problem—solicit information from them, and do not wait for the progress meeting. Allocate tasks in sufficient time for team members to plan their own work and check that the task requirements have been understood. If it is likely at any stage of any given task or subproject that the team member is struggling to achieve the target, set intermediate milestones (e.g., if a task is to complete the specification for the manufacturing facility in 12 weeks, include milestones of production process design in 4 weeks, create a *wish list* of desired equipment after 6 weeks, investigate the options during the next 4 weeks, and, during weeks 11 and 12, list the required equipment. This setting of *deliverables*, as they are sometimes called, maintains morale and ensures the task is completed on time. But beware of applying such close supervision on someone who would plan the 12 weeks quite differently and still achieve the target. Ensure that, in addition to managing the project, the needs of the team members are met.

To maintain motivation, create a sense of belonging though actions or attitude (e.g., commission mugs or pens with the project title that are issued only to team members, socialize together, and celebrate key targets with treats).

Communication and clear definition of the requirement is particularly important when dealing with external contractors. Any problems or delays always result in frustration and worry unless a clear specification document is issued. In many cases, expensive legal issues arise simply because the contract requirements are not clearly communicated. Check the contractor's reputation for technical competence, quality, delivery reliability, and credit rating. Invite discussion of the specification and, finally, agree in writing what will be carried out and when. This negotiation ensures that the supplier is in agreement and can be used in the event of contractual problems, but note that, in order for it to remain valid, no changes can be made to the specification without prior agreement.

Project management is one of those techniques that can be carried out adequately with simple common sense but inevitably takes some time to perfect for any company. However, it is always important to ensure the support of senior management and to avoid the need for crisis management!

13.6 THE LAUNCH OF A NEW PRODUCT

The introduction of the new technology into the organization has hopefully been based on both good innovation philosophy in the company and detailed planning on product development.

Some new technology may not be immediately visible to the customers; rather, it is apparent through intangible benefits, such as improved product quality, delivery times, or flexibility.

However, the new technology may be an opportunity to acquire good publicity or may require new marketing approaches. It may also allow, through publicity activities, a sense of ownership of and pride in the new development by all staff in the company.

The company probably already has well-thought-out policies on advertising and public relations, but the aim of this section is to emphasize good practice and clearly define often seemingly interchangeable terms such as public relations, publicity, advertising, sales promotion, and selling plans.

It might be that the company will opt for a *total communications solution* such as the service offered by the Dentsu advertising agency in Japan. Dentsu will provide the normal *above the line* (i.e., media) activities such as TV and print advertising, and *below the line* activities such as direct mail, special offers, trade fairs, sponsorships, conventions, and events. It will even act as custodian for the company's brand profile, covering everything from a design for headed paper to sales promotion and marketing. The company will also provide new product ideas, develop the ideas, create new business opportunities, manage business trips, and analyze research data or social/consumer trends.

Taking some key terms individually:

- *Public relations.* "Public relations practice is the planned and sustained effort to establish and maintain goodwill and mutual understanding between an organization and its publics" [3]. Thus, it is continuous rather than haphazard, it is based on objectives, and the results can be measured.
- *Publicity.* Publicity can be defined as the result of information becoming known by the general public (however, in the advertising/PR world, publicity is often loosely equated to advertising or PR!)
- *Advertising.* "Advertising presents the most persuasive possible selling message to the right prospects for the product or service at the lowest possible price" [3]. Thus, it is about selling (as compared with informing or creating understanding in PR). Advertising is more likely to succeed if PR activity has created knowledge and understanding of the product (so-called *market education*). The effectiveness of finding the right prospect depends on market research. A good advertising agency uses creative, research and media-buying skills to produce economically productive advertising.
- *Sales promotion.* This below-the-line concept consists of fixed-length schemes to launch a new product or revive/increase sales of an existing product. It can be short-term (e.g., money-back offers, competitions, free gifts, cash refunds) or long-term (e.g., regular product demonstrations). It is often used as an alternative to media advertising, and its costs are similar to those of advertising.

- *The selling plan.* The introduction of new technology and its effect on product development will require thought about possible changes to the selling plan described within the company's original business plan. For instance:
  - Who will be doing the selling and what changes to their training will be necessary?
  - Will the selling methods be changed (e.g., new point-of-sale material, new mail shots, cold calling, telephone calling)?
  - How will the sales effort be monitored and controlled?
  - What questions do you anticipate and what arguments do you have to meet these questions?

How will the market research on customers (as described in Chapter 3) be tied to the selling plan?

There are several differences between these concepts. PR is a much bigger activity than advertising or sales promotion. It relates to all communications of the company, whereas advertising/sales promotion are mainly based on the marketing function. PR is not free or unpaid-for advertising. PR is definitely not free. It is time consuming, and time costs money (often through consultants). Advertising may not be used by an organization, but every organization is involved in PR (e.g., the police force does not advertise its product but certainly engages in PR).

Sales promotion does bring the producer closer to the customer, and it would be foolish for a sales promotion to produce bad PR (e.g., the fiasco in 1994 when the U.K. Hoover Company was unable to meet the demand for free air tickets during a promotion). However, sales promotion is not PR as defined here.

All parts of the marketing mix as described in Chapter 12 (naming, packaging, research, pricing, selling, distribution, after-sales service) bear some degree of communication and goodwill.

13.7 CONCLUSIONS

After passing through the stages of the technology-transfer task list up to the stage where the go-ahead has been given, failure of the project can still occur due to poor operations management or a bad launch of the product/service.

Recently, a local businessman commented to one of the authors that it was fine for us to recommend to him the need for basic financial skills, an understanding of training, and a knowledge of information sources, but he was only human. One of the human traits he maintained was being either good or bad at managing projects (including launches). He said it was "genetic," like being good at golf or drawing. You can do it or you can't. He'd had psychometric testing, and he was an "innovator" rather than a "completer." We would disagree with such resignation to one's genetic fate. For managers responsible for technology transfer, basic project-management skills can be learned, checklists made, plans written, and communication routes organized.

Similarly, the launch is not just something that happens well to the gifted. It is part of the overall image of the company, where selling is just as important as production engineering or design. The launch can make or break a new technology-based product. A successful launch can bring additional benefits, such as an improvement of morale and internal communication.

References

[1]   CBI/DTI, *Innovation*, Company Reporting, Ltd., Edinburgh, Scotland, 1993.

[2]   "Reasons for Small Business Failure," London Business School, 1988.

[3]   Jefkins, F., *Public Relations*, 4th Ed., London: Pitman, 1992.

Selected Bibliography

Kuczmarski, T. D., *Managing New Products: The Power of Innovation*, 2nd ed., London: Prentice-Hall, 1992.

Chapter 14

Training and Human Resources

14.1 INTRODUCTION

Training is a fundamental part of the human resources management policy of any company. Human resources staff members (still often called *personnel* staff) should have the skills to design systematic training policies, plus they should be able to help the organization recognize that continuing development of staff is required after the initial or new development training is finished.

During the 1980s, the debate over the effect of new technologies in reducing employment and de-skilling workers gradually changed to discussions on more flexible requirements for skills. The level of vocational training varies greatly between nations; it is high in Germany, low in Britain, and it is higher in the Far East than it is in all of the EU nations.

Further problems come in some certain geographic areas (e.g., North Carolina, with its concentration of high-tech industries), where there is a high turnover rate of staff or where an uneven commitment to training leads to poaching from good trainers (recruitment rather than training). It has been said that if a company loses a lot of well-trained staff members to others, then its training policy is right but nothing else is. A turnover rate of 10% a year can bring heavy costs, which instead could be devoted to capital investment or training. Examples of these extra costs are:

- Increased supervision;
- Effects on quality control/reworking;
- Training;
- Human resources expenditure;
- Stress-related absences;
- Loss of continuity for customers/suppliers.

Employers can retain the loyalty of staff with the correct *retention strategy*. This should go beyond simply paying more. The very interesting process of Human Performance Engineering™ (HPE) [1] seeks to address the complexities that can determine the attitude of employees. HPE "creates new job descriptions and accountability review systems which are as dynamic and open to change as today's high-tech responsibilities" [1]. It is claimed that hiring, development, and performance critiques can be guided with the same language used on the production floor. Supervision and training can be reduced at the same time as morale and retention rates are increased.

Over all these issues lies the central theme that training is essential for competitiveness, growth, and innovation. This chapter will study four main areas—all of which contain the word *training*, but all are quite different in their purposes:

- Training of the whole organization to accept change and increase commitment;
- Management training and development;
- Training of staff;
- Customer training (preferably as part of the sales philosophy).

There appears to be a general industrial consensus that practices under these headings often have to be subdivided by the size category of the company.

14.2 TRAINING IN THE ORGANIZATION FOR INNOVATION AND CHANGE

Chapter 12 described the type of culture-change programs that can be introduced to bring about a new climate based on quality, customer service, innovation, technology transfer, and product development. All of these change programs had initial and ongoing training built in.

In the Goodyear turnaround, after the 12 objectives for the 1990s were determined, a large meeting of about 1,000 staff members was called. This was videotaped to show to staff around the world, regardless of position in the company. The ownership of the objectives by everyone was sought. Since then, the regular status meetings on the company have been taped and circulated. The introduction of cross-functional teams was also backed by extensive training in team-building/communication skills. This will be seen in Section 14.3 when team building is considered in more detail.

Commitment to an organization has three main factors :

1. A strong desire to remain a member of the organization;
2. A strong belief in the values and goals of the organization;
3. A readiness to exert considerable effort on behalf of the organization.

Commitment can be increased but there are necessary reservations about this:

- It has to be accepted that the interests of the organization and the staff do not necessarily always coincide. Any developments by the organization should clearly spell out the implications for staff and the justifications.
- Values should be clear to staff but management should allow and encourage flexibility within responsible teamwork and well-defined roles/appraisal.
- Senior management should not expect too much from commitment programs. They can increase job satisfaction, loyalty, and retention of staff, but this will only produce increased productivity if other quality planning has also taken place.

The late 1980s produced *the learning organization*. The learning organization does the following:

- Encourages people to identify their own learning needs;
- Provides individuals with regular reviews of learning and performance needs;
- Provides feedback on performance and achieved learning;
- Provides new experiences from which people can learn;
- Facilitates the use of training on the job.

14.3 MANAGEMENT DEVELOPMENT AND TRAINING

At the most senior level, the main need is to clearly articulate the company's mission and to demonstrate actual commitment to this mission. TQM training should be widely available to senior staff wishing to sharpen their recognition of their leadership role.

However, it must be recognized that corporate managers who grew up in the industrial "bureaucratic" world will often show the characteristics of the "expert," whether it is in the area of finance, engineering, research science, or information technology. Such experts may find teamwork or cross-function groups uncomfortable. A paradigm of reward based on individual/section performance has to be changed to demonstrate reward for team performance across the organization. Intense support will be required for the individual to want to overcome long-learned personal traits. AT&T (advised by the Maccony Group of consultants, leaders in this field) invited employees' spouses to "effectiveness workshops" to encourage the new habits in the household as well as in the company.

At the middle-manager level (assuming the 1990s organizational structure retains the concept of middle manager in a particular company), management development has three main objectives:

1. Improving the performance of managers by seeing that they are clearly informed about their responsibilities and by agreeing on objectives by which performance can be appraised with them.
2. Identifying managers with further potential and ensuring that they receive the required development, training, and experience.
3. Ensuring that succession within the organization is adequately planned and monitored.

Formal approaches to management development include:

• Development on the job;
• Development through work experience;
• Formal training;
• Structured self-development.

A typical list of management competencies for a company considering change through innovation and technology transfer are:

• Intellectual:
  • Strategic perspective;
  • Analysis and judgment;
  • Planning and organization;
• Interpersonal:
  • Coaching/teaching but not abdicating;
  • Persuasiveness;
  • Assertiveness and decisiveness;
  • Interpersonal sensitivity;
  • Oral/written communication;
• Adaptability and resilience;
• Results orientation:
  • Energy and initiative;
  • Achievement motivation;
  • Business sense.

These core competencies should be backed by an understanding of scientific/technological working methods, mentalities, and needs (e.g., fear of remote company management, some freedom to experiment, some thinking time).

Training should be considered on tools for stimulating creativity such as:

• Morphological analysis;
• Brainstorming;
• Pareto charts;

- Ishikawa, or cause-and-effect diagrams;
- Questionnaires;
- Suggestion schemes;
- Scenarios.

14.4 STAFF TRAINING

Staff training is characterized in many companies by a further division into:

- Graduates/researchers;
- Intermediate/technician staff;
- Blue collar/retail staff.

While flexible TQM-based training should build bridges across both functions and "grades," there still exist issues such as:

- Shortage of skilled graduates;
- The rift between graduates/researchers and aspirational technician/intermediate staff;
- Geographic-based shortage of some skills;
- The decline of apprentice processes;
- Lingering unwillingness of specialist staff to be associated with the less prestigious image of production;
- The difficulty that some staff members have at coping with change.

In the Baden-Württenberg area of Germany, Bosch is a good example of a company with a serious commitment to training. The Bosch college was set up in 1981 to keep college-educated staff updated on changes in scientific, technological, and economic knowledge. Although the use of apprenticeship is still strong, extra budgets have had to be set aside to cope with the need for "further days" of updating on new multiskills.

At the other end of the company size scale, Renishaw Transducer Systems of Gloucestershire, England, has shown how a company with 50 staff members can deal with some of these problems. Renishaw found that physics graduates required approximately three years of internal training both on and off the job before a significant contribution was made. With sponsored students, who spent one year at the company before going to college (plus their vacations), there was a much more rapid payback. Recently, Renishaw has improved this payback period by allowing new graduates to move between the design, production, and marketing departments in their first two or three years. Developments such as a new laser cutting tool and a ballbar have come to market up to 70% more quickly because of the cross-functional work of the new graduates. But Renishaw also appreciates the skills and flexibility of technical staff, who are encouraged to undertake degree courses part time. Some of the first of these new part-time route graduates now run the production test and software design departments. In addition, there is considerable contact with end users during the training.

The nature of the product itself can affect training needs due to cyclical changes. The development of the microelectronics industry provides an excellent example of this. The requirement in the 1990s has been for hybrid skills geared toward integrating different parts of a system rather than specialist skills geared toward concentrating on the parts. These changed skills come from a need to match customer expectations on extra features and on quality. While the 1980s concept of the *flexible firm* did not turn into widespread reality, some form of flexible/hybrid skilling is necessary to meet competitive situations. Flexibility comes from both competency in the workplace and the full utilization of staff capabilities.

A key concept of flexible working or cross-functional product development is teamworking. This can range from the large platform teams into which Chrysler was reorganized to cross-functional groupings of three or four people in smaller companies. Wide experience, especially in the U.S., has shown that the training required for teamworking is usually underestimated. It should have the following elements:

- Managerial, leadership, and supervisory training, which should take place before the first teams are established. American practitioners have widely used at least 10 two-hour sessions over two or three weeks. Subjects might include team building, empowerment, communications, problem solving, motivation, quality, and participative management.
- Operative training, which falls into three main areas:
  - Team working and related skills, which often prove very difficult to achieve fully;
  - Technical issues, whereby team members need to know more about specific areas, such as machine maintenance, quality, line balancing and costing;
  - Multiskilling, whereby a strategy is decided to gradually increase the ability of each member of the team to carry out all of the operations in the team. Multiskilling can be achieved by conventional training methods, but it can also be achieved by one member of the team with a particular skill training two or three of his or her colleagues. This also helps raise the self-esteem of the individual doing the training and helps develop team communication and cohesiveness.

14.5 CUSTOMER SERVICE TRAINING

When selling a system rather than a list of products, the training of customer service staff can be an important factor contributing to the overall package.

Conventionally, the training aspect of the order is met by printed training manuals and the provision of onsite training. However, an awareness of other methods of customer service staff training can lead to a more attractive bid package and a longer-term, safer use of the new system.

A good example of this was a U.S. contract won in 1992 by Simon Holder of Bury, England. The contract was for a fully automated processing plant for the recycling of corrugated containers in Jacksonville, Florida. Working with the consultants YPL Communications, the Simon Holder firm studied the customer service staff requirements in detail. These staff members would have mixed skills and learning abilities, and the staff turnover rate would be quite high. Also, there were language and associated cultural considerations in an area with a high percentage of Hispanic people.

For an outlay of $100,000 out of a total contract of $102 million, the answer proved to be an interactive video training system in which the operation of the entire plant is converted into a series of audio-visual programs combining video footage with stills, graphic explanation (including some computer animation of key processes), and first-person commentaries. The entire program takes the form of 20 laser discs held at central training points within the new complex. These are played on a laser disc player linked to an ordinary TV set that has one notable extra feature in the shape of a bar-code indexing system. Staff members wishing to learn about any section of the operation insert the appropriate disc and, using the bar-code reading wand, pick the appropriate subsection from the bar-coded contents lists in plant operations and maintenance workbooks or follow on-screen indexing prompts.

The program was felt to be ideal in view of the mixed abilities of the staff involved. Viewing can be repeated until the instructions are mastered. In addition, of course, the laser discs remain when conventional onsite training would have finished. Simon Holder felt that the innovativeness of the system played a major part in its bid. As a competitive resource, the ability to produce programs of this kind has a cumulative effect in gaining new business. Each subsequent program can, where appropriate, draw on existing material and experience. The program-making process also copes well with different languages or soundtracks. Simon Holder used both a professional U.S. actor based in London and a Hispanic speaker from Florida in a demonstration program as part of its bid.

14.6 CONCLUSION

This chapter has stressed the importance of different types of training to deliver the message of innovation, technology transfer, and bringing the transfer to its successful implementation.

Evaluation of corporate and staff attitudes is important (as part of the technology audit) to establish a starting point. For example:

- Should a major culture change be implemented?
- Are creativity and technology recognized as important by management and staff?
- Are the mechanisms for technology transfer understood?
- What gaps exist in knowledge and skills?
- Is training genuinely ongoing and not just a one-time fix?

References

[1]   Stillwagon, W., and R. Burns, "Improving manufacturing competitiveness through the application of Human Performance Engineering^{TM}" *International Journal of Technology Management*, Vol. 8, Nos. 3/4/5, 1993, pp. 411–421.

Chapter 15

Control and Performance Measures

15.1 INTRODUCTION

Information on the progress of new technology projects is often an overlooked ingredient of success. A *management information system* (usually IT-controlled nowadays) is essential for monitoring performance against the business plan and budget targets. Measurement should be accurate and fast, but, very importantly, it should be easy to use and should reflect actual projects rather than being just a mass of unprioritized data.

With a well-designed management information system in place in the company, it is possible to have a similarly good monitoring and control system. Many technology projects have failed due to a lack of knowledge about the project's progress until it was too late.

15.2 THE LIMITATIONS OF EXISTING SYSTEMS

It has been found that traditional controls and performance measures (usually based on cost accounting) have severe limitations when applied in environments using new technology.

Examples of these limitations are :

- Efficiency measures based on output divided by input often actually decline in small-batch, flexible, or JIT situations.
- Great emphasis in the past has been put on maximizing output (with accompanying bonuses to workers). This, of course, can result in increased inventories and financial charges. It should be noted that new techniques like JIT, by reducing stocks, can show a one-time drop in internal profits (if

absorption costing is used to divide overheads). It is as well to be aware of the possibility of this one-time drop in advance.

- Standard costs for stages of operations are inappropriate in a rapidly responsive manufacturing situation. They can, for instance, been seen as targets, which goes against the philosophy of constant improvement.
- Labor costs often have overhead costs attached to them, causing a belief that reducing labor costs (rather than studying indirect costs) will produce major savings.
- Quality, reliability, design lead times, research/innovation effectiveness, customer satisfaction, and flexibility have often not been monitored in financial accounting–dominated systems.
- Feedback reports (being compiled weekly or monthly) have often arrived too late to be of value.

To overcome these limitations new measures can be introduced:

- Technology transfer/research/innovation effectiveness measures;
- Quality measures and control;
- Production/cost measures and control.

These three types of measure are discussed in the next three sections.

15.3 TECHNOLOGY TRANSFER/RESEARCH/INNOVATION EFFECTIVENESS MEASURES

Technology transfer, as we have seen throughout this book, can be a transfer within a company or from an outside source. Both of these approaches can have measures of performance, although the limitations of such measures must be understood. For instance, a tempting measure of internal research might be the number of scientific papers published by staff. However, further investigation of this measure reveals the difficulty of correlating scholarly activity with products launched, product success, and product profitability.

For technology transfer from external sources, possible measures are:

- Number of technology transfer projects currently underway;
- Number of licenses signed for external technology in the last year;
- Rating of the success of the new technology at meeting its intended requirements;
- Lead times in technology transfer–based products compared with nontechnology transfer–based projects;
- Rating of efficiency of company information scanning systems;

- Findings in the technology audit comparing technology-transfer practices with benchmarked competitors;
- Percentage of new products using technology developed outside the company;
- Percentage of sales due to products using technology developed outside the company;
- Profitability of products (as a percentage of all profits) due to products using technology developed outside the company;
- Rating of the degree of understanding within the company of the technology-transfer process;
- Rating of the degree of support within the company for the importance of technology transfer;
- Investment in technology transfer as a percentage of sales.

For the internal research/innovation process (and its transfer to the production process) measures might include:

- Performance of the company at selecting successful projects to pursue;
- Success of company at producing ultimately successful projects (i.e., on time, within budget, meeting specified requirements, accompanied by full documentation);
- Quality of experimental testing;
- Rate of generation of new ideas;
- Frequency of use of pilot plant;
- Rating of success of teamworking;
- Number of customer contacts per research staff;
- Rating of internal communication;
- Staff turnover rate;
- Rating of staff morale;
- Relevance of produced documentation to what is required;
- Documentation availability as a percentage of the number of occasions that documentation was required.

15.4 QUALITY MEASURES AND CONTROL

These can be divided into three categories:

1. *Incoming materials.* Some companies now rate their suppliers on a quality basis as well as on price and delivery. A measure based on price can affect the quality measure by encouraging cheaper, poorer quality materials.
2. *In-process controls.* These include establishing control points at various stages in the production process to track variations from specification at each stage or running quality-audit programs that randomly measure quality at specific

stages. Quantities of scrap, rework, and defects as a percentage of good product are obtained (and these quantities can then be costed).

3. *Customer satisfaction.* This is the ultimate measure of quality. It can be measured by passive means such as warranty claims, returns, allowances, and letters of complaint/commendation. More proactive companies survey their customers on a regular basis.

A checklist of quality measures for control might be:

- Product durability;
- Product reliability;
- Product performance;
- Conformance to tolerances;
- Range of features;
- Uniqueness of features;
- Appropriateness for intended use;
- Ease of use;
- Ease of maintenance of product;
- Appearance;
- Return rate of supplies;
- Customer complaint rates;
- Field repair rates under warranty;
- Use of certified suppliers as percentage of the number of suppliers and by volume of supplied material.

Oakland's book (see the Selected Bibliography at the end of this chapter) is highly recommended for its discussion of quality measures.

15.5 PRODUCTION/COST MEASURES AND CONTROL

In a new approach to measurement, output is measured at the point of sale (rather than in the finished goods stock). This maximizes what actually goes to the customer. Bonus schemes to workers can be bought out or changed to group schemes based on factory dispatches. Input measures are changed to include only direct productive time. In addition, measures of time in which a given number of parts can be produced are used to acknowledge the benefits of flexibility.

Delivery measures can be based on *total cycle time* (the length of time from the placing of an order by a customer to delivery) or *manufacturing cycle time* (the sum of processing time, inspection time, wait time, and move time). In the manufacturing cycle, only processing time adds value. The aim of any new technology transfer is to eliminate non-value-added activities. Reducing setup times leads to shorter manufacturing cycles and greater flexibility. Also, preventative maintenance is encouraged to ensure good quality and on-time deliveries. Machine down time is also a useful measure, provided that it measures time when the machine is actually needed.

Flexibility measures include total launch times for new products, delivery times, percentage of on-time deliveries, and the ability to change the production mix.

Innovation measures include an assortment of key characteristics of new products relative to those of competitors, as well as feedback on customer satisfaction with new products.

Inventory reduction is a key objective, and measures influencing inventory levels include supplier delivery performance, setup times, equipment breakdown, and quality.

New methods for cost control are being introduced based on activities rather than functions. These ensure the pinpointing and control of activities that cause costs (cost drivers) rather than products, which do not. The attraction of such an activity-based costing (ABC) system is that it provides a basis for identifying costs more accurately and highlighting the absorption of non-value-added overheads into often vague functional costs. However, while companies investing in ABC always use the new information in product-related decisions, many do not use it to systematically review non-value-added activity. They thereby miss the opportunity to fully gain the benefits of ABC. In the 1991 Price Waterhouse "UK Survey of Cost Management Techniques and Practices," 32% of respondents used ABC but only 11% were using cost drivers as performance measures [1].

Delivery reliability involves the supplying of products on the agreed date. On-time delivery involves both the manufacturing function and the distribution function. Very often, delivery performance is an order-losing performance quality measure. The exactness of the due date can vary from a slot of 20 minutes in a highly structured manufacturing schedule to delivery in an agreed week. Delivery-time data is an important source of management information.

A checklist of possible production/cost measures might be:

- Rejection rates;
- Proportion of plant-measurable processes under statistical control;
- Proportion of orders delivered on time;
- Production lead time as a proportion of total delivery lead time;
- Value of materials in the plant as a proportion of production total;
- Value of finished goods in the plant as a proportion of production total;
- Production down time as a percentage;
- Mean time for repair;
- Mean time between failures;
- Measures for safety.

15.6 CONCLUSION

Spectacular financial collapses like that of the Barings investment bank in 1995 are widely known examples of the need for effective information and control. But there are also numerous smaller examples in technology transfer of potentially good projects that fail to deliver the intended benefits due to a similar lack of monitoring and control. Barings made a broader point however on "effective" control. The bank had a world-class management information system and control mechanisms did exist, but senior management did not or could not use them.

This chapter has listed a wide variety of possible measures that might be considered for use. It is likely that you could not immediately put all of them into action, but try to consider them in the light of what you really need to know about the progress of the whole project from information gathering to final production and delivery. Even a few basic measures are much better than none.

References

[1] "U.K. Survey of Cost Management Techniques and Practices," Price Waterhouse, London, 1991.

Selected Bibliography

Oakland, J. S., *Total Quality Management,* 2nd ed., Oxford, England: Butterworth-Heinemann, 1993.

Conclusions

In this book we have looked at the factors that make up the process known as innovation and technology transfer. These factors are the parameters most influential in the successful implementation of a new product or service or successful change of product or service. At this stage, it is a good time to reflect on the company as a whole and review some of the important concepts in the book. It is worth noting again the most important features of an innovative company. These are as follows:

- A long-term orientation;
- A commitment to change from top management;
- Creativity and a responsiveness to new ideas;
- Identification, capture, and transfer of new knowledge;
- The presence and encouragement of internal entrepreneurs;
- Awareness to trends in the total business environment;
- A focus on user needs and receptivity to user ideas;
- A high level of communication, both internal and external;
- Flexibility to enable a rapid response;
- An external orientation;
- Cross-discipline integration;
- Strong strategic planning with progress monitoring;
- Investment in education and training to support the change.

In an organization, the creation of the proper environment for the generation and implementation of technical change is dependent upon a number of interrelating elements, namely:

- Corporate culture;
- Strategy formulation, dissemination, and feedback;

- Organizational structure;
- Managerial information and control systems;
- Attitudes, motivations, and contributions of individuals.

In his book *Innovation Strategy* [1], Alan West provides what he calls "the golden rules of innovation." We feel that they encapsulate many of the ideas in this book and are worth noting and keeping in mind for anyone striving for success in business. An adaptation of the golden rules is as follows:

- Think strategic:
  - Fit the resources of the organization to the conditions of the market.
  - Control the market and the competition.
- Think novel:
  - Develop approaches to products, services, and customers that are demonstrably different from those currently being offered.
  - Use novel approaches to extend and promote old concepts and products.
  - Ensure that no possible avenue for both current and future development is ignored.
- Think customer:
  - What does the customer want?
  - What are the benefits the organization can offer?
  - Define customer needs in all aspects of the product.
- Think detail:
  - Accurately define the components of each stage of the development process.
  - Create detailed specifications of the product or service.
  - Develop a comprehensive engineering analysis program.
  - Create a predetermined testing program that will identify the major potential problem areas.
  - Create the correct balance in skills and personality of personnel in teams.
- Think betterment:
  - Analyze the work requirements in each task.
  - Build on existing strengths.
  - Assess where investment would be most appropriate.
  - Concentrate on long-term sales.
  - Look for real growth in added value per employee rather than in total return on capital employed.
- Think people:
  - Create an enthusiastic and effective workforce.
  - Employ the right people in the right jobs.
  - Develop people.
  - Create a belief that changes are made to suit the workforce rather than forcing the workforce to change to suit the organization.

- Think adaptability:
  - Be more knowledgeable, more responsive, and quicker than the competitors.
  - Understand the customer.
  - Get change implemented without pain.
  - Produce more added value.
  - Act rather than reflect.
- Think future:
  - Plan ahead for contingencies.
  - Be aware of the need to change or adapt.

All of these points can be considered of equal status. None can be singled out as being more important than the other. They form a very good checklist for any business person in virtually any sector to be aware of and to constantly review as an appraisal of his or her own procedures and development. An innovation program that addresses all of these issues and creates the correct environment is without doubt giving any organization the fullest chance of success. Conversely, a company that chooses to ignore the elements is pursuing the high road to probable disaster.

We do stress the very final point, to "think future," because as John Galsworthy wrote in *The Forsyte Saga* [2]:

"If you don't think about the future, you won't have one."

References

[1]   West, A., *Innovation Strategy*, Prentice-Hall, 1992.
[2]   Galsworthy, J., *The Forsyte Saga*, London: Oxford University Press, 1995.

Appendix A
EU Framework Programme 1994–1998

Research Areas	*Cost in Billions of ECUs*
Activity 1: Implementation of research, technological development, and demonstration programs by promoting cooperation with and between undertakings, research centers, and universities	10,686
1. Information technologies	1,932
2. Telematics	843
3. Communication technologies	630
4. Industrial and materials technologies	1,707
5. Measurement and testing	288
6. Environment and climate	852
7. Marine science and technology	228
8. Biotechnology	552
9. Biomedicine and health	336
10. Agriculture and fisheries	684
11. Non-nuclear energy	1,002
12. Nuclear safety	414
13. Controlled thermonuclear fusion	840
14. Transport	240
15. Socioeconomic research	138

Research Areas	Cost in Billions of ECUs
Activity 2: Cooperation with non-member countries and international organizations	540
Activity 3: Dissemination and application of research findings	330
Activity 4: Stimulation of the training and mobility of researchers	744
Total	*12,300*

Appendix B
Specific Programs of the
IV Framework Programme

B.1 ACTIVITY 1

B.1.1 Information Technologies

Aims: Specific research and technological development program in the field of information technologies.

Areas of research:

- Microelectronics;
- Information processing systems and software;
- Advanced business and home systems peripherals;
- Computer-integrated manufacturing and engineering;
- Basic research;
- Large-scale targeted projects.

B.1.2 Telematics Systems

Aims: To provide the conceptual, technological, and prenormative basis for the gradual implementation of trans-European telematic service networks.

Areas of research:

- Support for the establishment of trans-European networks between administrations—the European nervous system (ENS);

211

- Transport services (DRIVE);
- Health care (AIM);
- Flexible and distance learning (DELTA);
- Libraries;
- Linguistic research and engineering;
- Telematic system for rural areas (ORA).

B.1.3 Communications Technologies

Aims: To continue the development of integrated broadband communication (IBC).

Areas of research:

- ISDN and integrated broadband communications (IBC) R&D;
- Intelligence in networks/flexible communications resource management;
- Mobile and personal communications;
- Image and data communications;
- Integrated service technologies;
- Information security technologies;
- Advanced communication experiments;
- Test infrastructure and interworking.

B.1.4 Industrial and Materials Technologies

Aims: To contribute to the regeneration of European manufacturing industry by basic technical research, integration of new technologies by user industries and technology transfer; to help small and medium-sized undertakings become more involved in transnational research and develop materials for the future.

Areas of research:

- Synthetic materials ceramics, superconducting alloys;
- Composites and intelligent materials;
- Molecular engineering;
- Nanotechnologies;
- Waste recycling;
- Digital modeling;
- Simulation of prototypes;
- Mechatronics;
- Miniaturization of components;
- Factory of the future.

B.1.5 Measurements and Testing

Aims: To improve measurements, testing techniques, and chemical analysis where they are not sufficiently accurate and where the measurement methods are insufficient to satisfy the new challenges in industry in the monitoring of the environment, food quality, and health.

Areas of research:

- Support to regulations and directives;
- Sectoral testing problems;
- Common means of calibration for the EU;
- Development of new methods of measurements.

B.1.6 Environment and Climate

Aims: To develop the scientific knowledge and technical know-how the EU needs to carry out its role concerning the environment and climate.

Areas of research:

- Participation in programs for the exploration of global environmental change (ENRICH);
- Climate change and climate impacts;
- Global changes in atmospheric chemistry and biogeochemical cycles and their consequences for life on Earth;
- Technologies and engineering for the environment;
- Research on economic and social aspects of environment issues;
- Clean environmental monitoring technologies;
- Treatment of effluent;
- Management of natural hazards;
- Optimum energy use.

B.1.7 Marine Science and Technology

Aims: To contribute to the establishment of a scientific and technological basis for the exploration, exploitation, management, and protection of European coastal waters and the seas surrounding the EU member states.

Areas of research:

- Marine science;
- Coastal zone science and engineering;

- Marine technology;
- Supporting initiatives;
- Large-scale multidisciplinary projects.

B.1.8 Biotechnology

Aims: To reinforce basic biological knowledge as a common and integrated foundation needed for applications in agriculture, industry, health, nutrition, and the environment.

Areas of research:

- Plant molecular engineering;
- Biosciences and bioprocesses;
- Human genome program;
- Biological materials;
- Ecology and population biology.

B.1.9 Biomedicine and Health

Aims: To contribute to improving the effect of medical and health research and development in the member states, in particular by better coordination of the member states' research and development activities and application of the results through EU cooperation and a pooling of resources.

Areas of research:

- Development of coordinated research on prevention, care, and health systems;
- Major health problems and diseases of major socioeconomic impact;
- Human gene analysis;
- Neurological, mental, and immunological disorders;
- Research on biomedical ethics;
- European medical network.

B.1.10 Agriculture and Fisheries

Aims: To upgrade and diversify agriculture and agricultural products; to enhance the competitiveness of agricultural and agrifood undertakings; to contribute to better rural and forestry management; and to ensure proper protection for the environment.

Areas of research:

- Primary production in agriculture, horticulture, forestry, fisheries, and aquaculture;
- Inputs in agriculture, horticulture, forestry, fisheries, and aquaculture;
- Processing of biological raw materials from agriculture, horticulture, forestry, fisheries, and aquaculture;
- End use and products;
- Target-oriented projects.

B.1.11 Nonnuclear Energies

Aims: To contribute to the development of new energy options that are both economically viable and more environmentally safe, including energy-saving technologies.

Areas of research:

- Analysis of strategies and modeling;
- Minimum emission power production from fossil sources;
- Renewable energy sources;
- Geothermal energy and deep reservoir geology;
- Energy utilization and conservation;
- Impact of new energy forms on the social and economic fabric.

B.1.12 Nuclear Safety

Aims: To contribute to the safe use of nuclear energy, taking into consideration all aspects of nuclear power plants and the full production cycle.

Areas of research:

- Radiation protection;
- Human exposure to radiation and radioactivity;
- Effects of radiation exposure on man: assessment, prevention, and treatment;
- Risks and management of radiation exposure;
- Reactor safety;
- Accident progression analysis;
- Behavior and qualification of the containment system;
- Accident management and control;
- Storage of long-lived radionuclides.

B.1.13 Controlled Thermonuclear Fusion

Aims: Joint creation of safe, environmentally sound prototype reactors; to prepare industry for the construction of the Next Step device, which is the achievement of self-sustained thermonuclear burn of a deuterium-tritium plasma and its control during long pulse operation.

Areas of research:

- Next Step design;
- Next European Torus (NET);
- International Thermonuclear Experimental Reactor (ITER);
- Longer term technical developments (including low activation materials, reactor blanket modules, and reference design for an electricity producing fusion reactor);
- JET (establishment of reliable methods of plasma purity control and preparation for the final phase of JET with deuterium-tritium plasmas);
- Support program:
  - Scientific support of the Next Step and to JET;
  - Study on alternative lines in toroidal magnetic confinement;
  - Other approaches to controlled fusion.

B.1.14 Transport

Aims: The implementation of a coherent transport policy and preparation of the ground for the means of travel for the future.

Areas of research:

- The integration of each mode of transport into a coherent multimodel trans-European network;
- Optimization of networks.

B.1.15 Socioeconomic Research

Aims: To anticipate future requirements.

Areas of research:

- The evaluation of science and technology policy options (technology assessment and forecasting);
- Education and training;
- Integration in Europe and social exclusion phenomena.

B.2 ACTIVITY 2: COOPERATION WITH NONMEMBER COUNTRIES AND INTERNATIONAL ORGANIZATION

Aims: The promotion of scientific and technological cooperation with developing countries and with non-EU industrialized countries.

Areas of cooperation:

- Specific projects on topics such as the environment and energy conservation;
- The European Economic Area:
  - Stronger links with the European Cooperation in the Field of Scientific and Technical Research (COST) program, EUREKA, CERN, European Science Foundation, ESA, ESO, and EMBL;
- The Central and Eastern European countries and the countries of the former Soviet Union:
  - Projects in nuclear safety, radiation protection, and radioactive waste management;
  - Assistance for the Economic Restructuring in Central and Eastern Europe (PHARE) program;
  - The Technical Assistance to the Commonwealth of Independent States and Georgia (TACIS) program;
  - The Joint Venture PHARE Programme (JOPP);
- Industrialized non-EU countries:
  - Cooperation with the U.S. in biology, thermonuclear fusion, molecular biology, and neurosciences;
- The developing countries:
  - Particularly in communicable diseases, desertification, improved crop yields, development of fishing and fish farming, and environmental protection.

B.3 ACTIVITY 3: DISSEMINATION AND APPLICATION OF RESEARCH FINDINGS

Aims: To promote the dissemination and exploitation of EU R&D results (SPRINT/VALUE program).

Activities:

- Research-industry interface;
  - New channels of information (CORDIS);
  - Utilization of results;
  - Promotional activities;
  - Identification of partners;

- Identification of financial support;
- Interface between research and the scientific community;
  - General context of research;
  - Communication of research;
  - Economics of research;
  - Management of research;
- Interface between research and society;
  - Contribution to the assessment of the social impact of science and technology;
  - Communication with the public;
  - Analysis of public demand and new requirements.

B.4　ACTIVITY 4: STIMULATION OF THE TRAINING AND MOBILITY OF RESEARCHERS

Aims: To help increase the human resources available for research and technological development, which will be needed by the member states in the coming years. To assist in the mobility of researchers and the formation of networks.

Activities:

- The development of an EU system of research training;
- Assistance for the creation and development of research networks for scientific and technical cooperation;
- The establishment of access to large-scale scientific and technical facilities;
- Euro-conferences;
- Sectors include all scientific and technological categories: exact, natural, social, economic, and human sciences.

Appendix C
Examples of Past and Present EU Programs

C.1 INTERNATIONAL SCIENTIFIC COOPERATION

These activities promote cooperation between laboratories in the EU and third-world countries in Latin America, Asia, and the Mediterranean. The cooperation takes place within the framework of bilateral cooperation agreements, which the EU has signed with the different countries.

The following countries are currently involved: Argentina, Bahrain, Bangladesh, Bolivia, Brazil, Brunei, China, Chile, Columbia, Costa Rica, Ecuador, El Salvador, Guatemala, Honduras, India, Indonesia, Israel, Kuwait, Malaysia, Mexico, Mongolia, Nicaragua, Pakistan, Panama, Peru, Philippines, Qatar, Saudi Arabia, Singapore, Sri Lanka, Thailand, United Arab Emirates, Uruguay, Venezuela, and Yemen.

The subjects include: medical research, biology, agriculture, environment, Earth sciences, material sciences, physics, chemistry, mathematics, and engineering.

C.2 EUROPEAN ACTION SCHEME FOR THE MOBILITY OF UNIVERSITY STUDENTS (ERASMUS)

ERASMUS promotes the mobility of university students in the EU to complete a recognized part of their course of study in another member state of the Union.

C.3 COMMUNITY ACTION PROGRAMME IN EDUCATION AND TRAINING FOR TECHNOLOGY (COMETT)

This program promotes cooperation between universities and industry regarding training in advanced technology, especially in response to technological change.

The program is aimed at people who have completed their initial training and those in active employment.

The following activities are supported:

- Cross-border interchange of university students, graduates, scientists, and specialists;
- Projects to promote continuing training in the technology sector and multimedia distance education;
- Supplementary and associated initiatives.

C.4 LINGUA

LINGUA is a program for the promotion of foreign-language knowledge in the EU. LINGUA seeks to improve foreign-language teaching and the learning of foreign languages within the Union. The program is aimed at:

- Language teachers who wish to deepen their professional competence;
- Universities and institutions that educate language teachers;
- Students who are studying foreign languages;
- Institutions that offer foreign-language courses for the business and vocational sector;
- Institutions that promote educational initiatives and exchange schemes for young people between the ages of 16 and 25 (especially as a part of vocational training).

C.5 TRANS-EUROPEAN MOBILITY SCHEME FOR UNIVERSITY STUDIES (TEMPUS)

This program is a trans-European mobility scheme for university studies for Central and Eastern European countries. It covers universities that specialize in the training field in Central and Eastern European countries. TEMPUS currently covers Poland, Hungary, Czechoslovakia, Bulgaria, Romania, Albania, Estonia, Lithuania, and Latvia.

The following activities are supported:

- Joint European projects linking universities and/or enterprises in Central and Eastern European countries with partners in the EU. Such projects may be linked, as appropriate, to existing networks, notably those funded in the framework of the ERASMUS, COMETT, and LINGUA programs.
- Mobility grants for teachers/trainers, students/trainees, and administrators.
- Complementary activities as exchanges of young people and activities of information.

C.6 ACTION FOR COOPERATION IN THE FIELD OF ECONOMICS (ACE)

The aim of this program is to stimulate exchanges involving economic techniques and concepts among economists in the Union and in the countries of Central and Eastern Europe. The research topics should relate to the process of economic reform and integration (microeconomic liberalization and reform, macroeconomic imbalances, and economic integration).

The following activities are supported:

- Research visits by economists and doctoral students from the above-mentioned countries to a university or research center in the Union;
- High-level seminars for economists and senior officials from the above-mentioned countries;
- Multinational networks and research projects involving at least one partner from the above-mentioned countries and two universities or research institutes in the Union;
- Professional and academic economists from the above-mentioned countries to attend economics conferences in the EU;
- Grants to institutions offering Ph.D. and MBA programs in the above-mentioned countries.

C.7 THERMIE

THERMIE supports projects for the application of new energy technologies whose realization is associated with a considerable degree of risk. The program is aimed at the following areas:

- *Rational use of energy:* energy savings in buildings and industry, more effective methods of heat and electricity generation and distribution, and energy savings in transportation and urban infrastructure;
- *Renewable energy sources:* solar energy, energy from biomass and waste, geothermal energy, hydroelectric energy, and wind energy;
- *Solid fuels:* environmentally benign techniques for exploiting coal and other solid fuels;
- *Hydrocarbons:* exploration and exploitation of oil and gas fields, transportation, and storage of oil and gas.

Research projects in these and related areas can currently be supported within the framework of the nonnuclear energies program.

C.8 DEMONSTRATION PROJECTS IN THE STEEL INDUSTRY

Support is given to projects aimed at promoting the commercial exploitation of new, high-performance, cost-effective techniques and procedures in the steel industry. Projects must be of an innovative character and must require levels of investment that ensure their technical and commercial feasibility.

Special attention is paid to projects aimed at reducing production and operating costs, improving product quality, providing support for steel sales, and adapting production conditions to environmental-protection needs. The program complements the Union's steel research initiatives, which have similar guidelines.

C.9 STRATEGIC PROGRAMME FOR INNOVATION AND TECHNOLOGY TRANSFER (SPRINT)

SPRINT is aimed at promoting the dissemination of new technologies and innovations by integrating national innovation infrastructures into a European network. This is achieved by promoting specific projects which are of particular demonstration value for innovative technologies, by coordinating national innovation promotion policies, and, in association with this, carrying out innovation monitoring.

The program is aimed at public and private consultancy bodies that, together with partners from other EU member states and on behalf of small and medium-sized enterprises, research and set up cross-border, technology-oriented structures of cooperation (especially technology-transfer projects).

Participants include: chambers of commerce and industry, technology and management consultants, innovation consultancies, consulting engineers, technology parks, and companies that carry out research on a contract basis.

C.10 VALUE

The aim of this program is to promote the dissemination and exploitation of EU research and development results, to help contractors identify partners and opportunities for financial support of project implementation, and to enhance collaboration and the transfer of innovation technology across the member states of the EU. The VALUE relay service is a dedicated network of local proactive advisory centers. The service is available prior to participation in EU-specific programs and throughout the duration of a research project, from start to final exploitation of results.

VALUE has been merged with the SPRINT program in the Fourth Framework proposals.

C.11 SCIENCE AND TECHNOLOGY FOR REGIONAL INNOVATION AND DEVELOPMENT IN EUROPE (STRIDE)

The aim of this program is to strengthen the research, technological, and innovative capacity of the regions whose development is lagging behind so that they are better placed to attract or retain highly qualified staff and technologically advanced activities in the productive sectors of the regions.

STRIDE also has a role to play in less industrially developed regions by promoting innovation for the diversification of local business and industry.

The measures of STRIDE include:

* Strengthening research facilities in objective 1 regions;
* Promoting participation and greater interconnection in EU and other international research programs and networks;
* Promoting linkages between research centers and industry.

C.12 INFORMATION MARKET POLICY ACTIONS (IMPACT)

The aim of this program is to establish a single market in information services. IMPACT aims to identify the strengths and weaknesses of existing information services in the Union and to stimulate and reinforce the competitive capability of European suppliers of information services. Other goals are to promote the use of advanced information services; to reinforce European cooperation to achieve a Union information services policy, paying particular attention to SMEs, the differences between the various regions, and the development lag of the less favored regions in the Union; and to make use of the results supplied by the other Union or national programs with a view to strengthening the information services market.

C.13 TRADE ELECTRONIC DATA INTERCHANGE SYSTEM (TEDIS II)

The aim of TEDIS II is to coordinate the development of electronic data interchange systems for trade, industry, and administration. The specific needs of the users, including those in small and medium-sized enterprises, must be taken into account.

The program will be carried out in coordination with other EU policies and initiatives (e.g., in information and communications technologies and in the security of information systems and standardization in the information market).

About the Authors

Ian Cooke graduated with a first-class honors degree in mathematics from the University of London. He was awarded his doctorate by the University of London for research in topology. He also gained a master's degree in electronic control engineering from the University of Salford, U.K. He has been awarded the statuses of chartered mathematician and chartered engineer. His business experience includes ten years of working as an information technology consultant in a wide range of organizations and industries. Dr. Cooke is a staff tutor and lecturer in mathematics, computing, and technology at the Open University, U.K. He can be contacted via I.E.COOKE@OPEN.AC.UK.

Paul Mayes is currently Head of Academic Information Services at the University of Teeside, Middlesborough, U.K. He has previously worked for Liverpool John Moores University and British Telecom. His research interests in technology transfer and the practicalities of quality management have led him to participate in the EU LINGUA, ERASMUS, DELTA, and COMETT projects, and collaborate with U.S. and European organizations. He can be contacted via PAUL.MAYES@TEES.AC.UK.

Index

www.ingramcontent.com/pod-product-compliance
Lightning Source LLC
Chambersburg PA
CBHW021430180326
41458CB00001B/201